CREATIVE
TOY TRAIN
TRACK PLANS

NEIL BESOUGLOFF

KALMBACH
BOOKS

Dedication

This book is dedicated to my wife Susy and our boys Ricky, Michael, Austin, Hunter, and Sammy. All of them graciously allowed me to monopolize our home computer night after night, week after week, as I clicked and moved hundreds of sections of track over and over to produce the 16 plans in this book.

Acknowledgments

Kent Johnson, former senior acquisitions editor in Kalmbach Publishing Co.'s books department and now an associate editor at *CTT*, came up with the idea for this project, gave it structure, and, most importantly, kept me motivated. And many thanks to Russell Becker of R&S Enterprises, who allowed us to include the 3D views of these 16 track plans.

This product is a Print on Demand format of the original book published by Kalmbach Publishing Company.

05 06 07 08 09 10 11 12 13 14   10 9 8 7 6 5 4 3 2 1

Visit our website at
http://kalmbachbooks.com
Secure online ordering available

ISBN 978-0-89778-530-3
Publisher's Cataloging-In-Publication Data
(Prepared by The Donohue Group, Inc.)

Besougloff, Neil.
 Creative toy train track plans / Neil Besougloff.

  p. : ill. ; cm.
 ISBN: 0-89778-530-4

1. Railroads--Models--Design and construction--Handbooks, manuals, etc.
I. Title.

TF197 .B47 2005
625.19

Managing Art Director: Mike Soliday
Book design: Sabine Beaupré
Track plan illustrations: Neil Besougloff and Terri Field

Contents

4 **INTRODUCTION**
Beyond sheets of plywood; toy train track lingo

9 **LAYOUT ONE**
Bridge out!
There's a real challenge on this
12- by 12-foot plan

13 **LAYOUT TWO**
Cold Mountain
A figure-eight track plan split in half
by a mountain range

17 **LAYOUT THREE**
Small Town
A 9- by 12-foot walk-in plan
set in middle America

21 **LAYOUT FOUR**
Old King Coal
Four coal-handling accessories
and three train routes

25 **LAYOUT FIVE**
Big City
Two trains on two loops
in the New York metro area

29 **LAYOUT SIX**
Round the Mountain
A winding reverse loop through
rugged western terrain

33 **LAYOUT SEVEN**
Prairiewood
Grain elevators, farm fields,
and three track loops

38 **LAYOUT EIGHT**
Breaking the Rule of One
A 12- by 12-foot plan with pairs
of operating accessories

43 **LAYOUT NINE**
The Quiet One
Walk-in, single-level operation in 9 by 12 feet

47 **LAYOUT TEN**
Through the Woods
Less is more on this 12- by 12-foot
walkaround plan

52 **LAYOUT ELEVEN**
The Works
Two intersecting loops and Lionel's
biggest tinplate accessories

57 **LAYOUT TWELVE**
Crossing the Ravine
An L-shaped 9- by 12-foot plan with
a dominating central bridge

61 **LAYOUT THIRTEEN**
The Transcontinental
Up and down the O gauge mountains
on the way to Promontory Point

66 **LAYOUT FOURTEEN**
A Day at the Races
Two intertwined ovals and six bridges
mean big-time operation

71 **LAYOUT FIFTEEN**
A Twist and a Flame
A two-train plan with a twisted loop
and fire-related accessories

75 **LAYOUT SIXTEEN**
A Tale of Two Cities
A continuous-run 12- by 12-foot
track plan with two destinations

80 **CONCLUSION**
Closing thoughts; list of manufacturers

Introduction

Inspiration strikes when you least expect it, and that bolt of enlightenment can originate from the unlikeliest of places. How else can you explain the unusual mix of origins for some of the O gauge track plans in this book: a rental car ride, oatmeal, singer John Mellencamp, a trip to Yosemite National Park, a missing bridge, New York City slang, a pizza with anchovies, and a book about American history?

That's no Lionel uncoupler track section to the left of a Burlington Northern Santa Fe freight train roaring through a cut near Woodward, Okla. It's an automatic flange oiler. *Hal Miller*

The 16 track plans in this book are a single person's vision. Use them as a whole or in part, or as inspiration for your own personalized layout. Track brands can be swapped, operating themes can be altered, and scenery suggestions can be turned upside down. It's all up to you—so open your mind and let the ideas flow.

Beyond boards

How many boards do you have? If you know what that question means, then you've been fully indoctrinated in toy train layout theory, circa 1975. "How many boards" means how many sheets of plywood. Often O gauge layouts are built in multiples of plywood sheets—4 by 8 feet, 8 by 8 feet, 8 by 12 feet, and so on.

"Boards" are mounted on sturdy legs, and, before track is laid and scenery constructed, a casual observer might mistake a budding Plywood Central for a Ping Pong table, an elaborate workshop bench, a laundry-folding table, or anything but a model railroad.

That was the world of toy trains 30 years ago. Today, with contemporary power tools, modern fasteners and adhesives, and natural and plastic foam-based scenery materials, toy train layouts don't have to be ovals of track defined by the dimensions of sheets of plywood.

All 16 of these track plans creatively break out of that rectangular world, but without leaving toy train traditions by the wayside. They share the following notable characteristics:

First, each plan is comprised of sectional track—either Lionel or K-Line tubular track, MTH RealTrax with built-in roadbed, Atlas O solid-rail track, or time-honored hi-rail styles, such as the pieces made by GarGraves and Ross Custom Switches.

Second, each plan readily accommodates postwar and modern operating and stationary accessories. They focus, however, on the core of the toy train hobby: continuous running of one or more trains.

Third, each plan is reasonably sized. Some fit in the confines of a 9- by 12-foot space, others within a 12- by 12-foot area. Some feature grades with over-and-under operation, while others are entirely on a single plane. Some plans are designed to walk around, others are designed to walk into, and still others are designed to fit against a wall or into a corner.

Fourth, each plan is asymmetrical. Although your ninth-grade geometry teacher might not approve, asymmetrical designs offer more interest than precise geometric loops.

Fifth, each plan can be wired for conventional-transformer control or be operated with Lionel's TrainMaster Command Control or MTH's Digital Command System. The choice is yours.

Sixth, each plan has a tangible connection to a real-life railroad past and present, or, to be more precise, a hobbyist's vision of a real-life railroad. None of the plans models a specific place on a map, but all present plausible and visually captivating locations for O gauge trains, track, accessories, and scenery.

Your reverse loop has a spur

My what has a who? Before jumping in, you should brush up on the lingo of toy train track. Some of these terms are familiar, but even seasoned hobbyists benefit from a full understanding of track terminology and how components work together to create an inspiring layout.

Here are some terms you'll hear in discussion of track:

Types and brands of track: Three-rail O gauge track falls into two major categories, which are defined by the construction of the rails.

Traditional Lionel O gauge track is called

"tubular track." In profile, the rails of Lionel O gauge track, made from sheet metal, are bent into a tube shape, hence the name. Lionel O-27 track, K-Line track, and Williams track also use hollow tubular construction. So do GarGraves, Curtis, and Ross Custom, but those three companies use rails with a smaller, more rectangular profile than the rails on traditional Lionel track.

Lionel's new track system, called FasTrack, also is made from sheet metal. However, instead of forming a hollow tube, its rails are molded around a raised plastic rib on top of the track's plastic base.

Atlas O track is an example of solid-rail track. The rails are formed from a solid piece of metal, either steel or a mixture of nickel and silver. MTH's ScaleTrax uses solid rails, and MTH's RealTrax, packaged in its train sets, uses solid rails affixed to a plastic road base. The base looks similar to the roadbed on Lionel's FasTrack.

Straights and curves: O gauge track is generally sold in straight and curved sections. Straight sections are generally 10 inches long, depending on the brand. Longer straight sections, upwards of 40 inches, are available in some brands. Track manufacturers also sell half-length straight sections and sometimes smaller "fitter" sections in lengths of 1 to 5 inches.

Curved track comes in a variety of diameters, all labeled with the letter "O" and a number, such as O-31 or O-72. These alphanumeric designations have been around since the 1930s. In the years since, track manufacturers have chosen the same designations for their track. An O-31 curve

Here's a creative mix of bridge styles that can be incorporated into a toy train track plan: a below-deck truss span meets masonry arches on the CSX in Maryland. *Stephen Panopoulos*

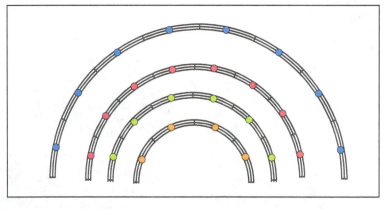

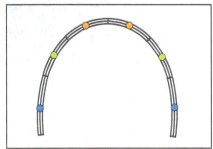

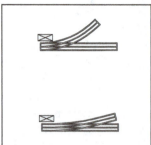

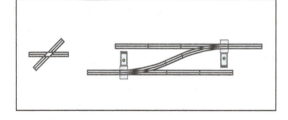

Top: Toy train sectional track is sold in a variety of curve sizes. The most common are O-31 (orange), O-42 (green), O-54 (red), and O-72 (blue).

Middle left: This is an example of an easement, using O-72 (blue), O-42 (green), and O-31 (orange) track sections.

Middle right: Track switches can have sharply diverging routes, like an O-31 switch (top), or feature gentle curves, like an O-72 switch.

Bottom: A crossing section—it looks like an "X"—and a crossover, which is comprised of two back-to-back track switches.

creates a circle of track 31 inches in diameter. An O-72 curve creates a circle of track 72 inches, or 6 feet, in diameter.

Ah, if it were just that simple, but it isn't.

Some curves form circles that are measured from outer edge to outer edge, while other measurements are taken from center rail to center rail. Blame it on tradition and the lack of a need for pinpoint accuracy in the early days of O gauge trains.

Today, common curve designations are O-31, O-42, O-54, and O-72. However, some manufacturers produce curves that are O-36, O-45, and O-63.

To add to the confusion, there's Lionel's O-27 tubular track. Originally available in 27-inch-diameter curves only, O-27 track, with a lower rail height than regular Lionel O gauge track, can today be purchased in other diameters, like O-42 and O-54. So yes, you can build a layout with O-27 track that has no 27-inch-diameter curves.

Curve easements: Some of the layouts in this book feature curve easements. Easements are used to soften the sharp

transition from straight-track sections to curved sections. For example, a straight section of track leading to one O-72 curved section, then one O-54 curved section, and then several O-42 curved sections would be an example of an easement.

Easements reduce the chance of derailments and, on a layout, look better. On the downside, they increase the amount of space needed to create a curve on a layout, and they also complicate the geometry of sectional track.

Switches and turnouts: Most O gauge track switches, also called "turnouts," mimic the curve diameters of O gauge track. An O-31 switch offers the same degree of curvature as an O-31 curved section; an O-72 switch matches an O-72 curve; and so on.

Keep in mind, though, that all manufacturers do not offer all sizes of curves as switches. You don't want to find out that, after purchasing several armloads of track, your brand doesn't offer an O-54 switch, yet your layout plan calls for one.

Some manufacturers, such as Atlas O, Curtis, and Ross Custom, also offer switches labeled by numbers instead of curve diameters. The diverging track on these switches branches off in a straight line and not an arc.

The numbers—such as "4," "6," and "8"—refer to the angle of the diverging track. On a no. 6 switch, it takes 6 units of measurement from the start of the diverging route before the two routes are 1 unit apart. Following this train of thought, the diverging route of a no. 4 switch angles off more sharply than the diverging route of a no. 8 switch.

Numbered track switches are favored by scale model railroaders and some hi-railers. For toy train layouts, the numbered switches are especially useful in yards, where you want switches that create closely spaced parallel tracks.

Crossings and crossovers: Crossings are the intersection of two rail lines. Crossings form an "X," and most track manufacturers currently offer 90-degree or 45-degree crossings.

A crossover, while conjuring up the same visual image as a crossing, is something else. A crossover fits between two parallel tracks. Also shaped like an "X," it allows a train on one track to "cross over" to the second track and vice versa. The effect is often achieved by using two sets of right/right or left/left track switches positioned with their diverging routes connected to each other.

Wyes: A wye is just what it sounds like— a track switch that resembles the letter "Y." Three wyes can be put together to form a triangle that can be used to change the direction of a locomotive. Think of making a K-turn back in high school driver's ed. class and you'll get the picture. A wye can also be created with regular track switches, although it would not be symmetrical.

Reverse or return loops: A key element to a successful track plan is a reverse loop. These loops allow a train running in a clockwise direction to make a 360-degree turn and return to the originating section of track but running in a counterclockwise direction. Ideally, a track plan has two reverse loops, allowing a train to change from clockwise operation to counterclockwise operation and vice versa.

Grades: Grades are sections of track rising or descending in elevation to cross over another track or an obstacle. The longer a grade the better, since that means a train has an easier climb to the top of a hill.

Fitter sections: Fitter sections, used to close irregularly sized gaps in a track plan, can be made or purchased depending on the style of track. Tubular track is easily cut with a saw, so you can make fitter sections as short as 1 inch. Solid-rail track, or track with built-in roadbed, is more difficult to cut, so manufacturers offer a variety of fitter sections as part of their lines.

Spurs and sidings: A spur is a length of track branching off a switch that ultimately reaches a dead end. If the end of a spur reconnects to the main line through another switch, then it's called a siding.

Yards: Yards are a series of spurs and sidings used to break down trains, sort cars, and reassemble trains. On toy train layouts, they also are ideal places for loading and unloading accessories.

Freight yards, while always visually interesting, require large amounts of space. That's why you won't see many large ones on the track plans in this book.

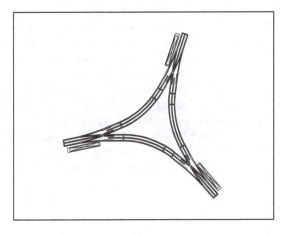

Wye switches can be used to form a triangle to turn around locomotives or even entire trains.

Here is an example of a reverse or return loop.

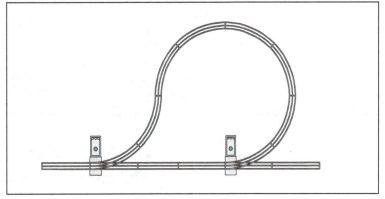

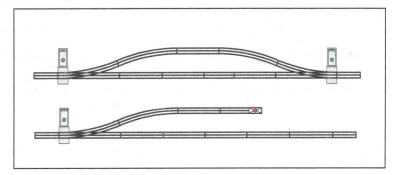

Point to point, out and back, and continuous run layouts: Trains on real railroads don't go in circles; they travel from one point or destination to another. Some scale model railroaders build extensive point-to-point layouts, but such plans are unsatisfying for most toy train enthusiasts.

An out-and-back layout is a point-to-point layout with a turnaround loop at one end but not at the other.

The most basic continuous-run layout is an oval of track, like those that come with train sets. All the layouts in this book are continuous run.

Track-planning tools

You can design a track plan with just pencil and paper, but once you use track-

The top drawing shows a siding. When a secondary track stops at a dead end, however, it is called a spur.

planning tools you'll wonder how you managed without them. CTT Inc. (not the magazine) offers a plastic track-planning template to keep your straights, curves, and switches in proper proportion as you create your design. The template looks somewhat like a stencil, and its use is intuitive.

The track plans in this book were initially created using "RR-Track" software (version 4.2) from R&S Enterprises. The software lets you choose a size for your layout, create its shape (whether rectangular or irregular), snap together track sections from a variety of O gauge manufacturers, and then add buildings, bridges, and landscaping, all appropriately scaled.

RR-Track also offers a three-dimensional viewing mode that allows you to virtually walk around your would-be layout, scenery and all, as if it really existed. A portion of

each layout in this book is shown in 3-D mode. If you're serious about drawing track plans, such software is well worth the investment.

The RR-Track software is designed for the Windows operating system. It can, however, be used on a Mac, provided you have translation software such as Virtual PC.

There are other track-planning software packages, such as "3rd PlanIt" from El Dorado Software, "3-D Railroad Concept and Design" from Abracadata Software, and "Cadrail" from Sandia Software. In addition, MTH and Atlas O offer free track-planning software, but only for their own track products.

Now that you're up to speed on track planning, turn the page and take a look at the first of 16 creative toy train track plans.

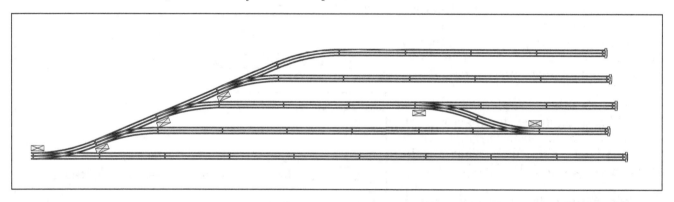

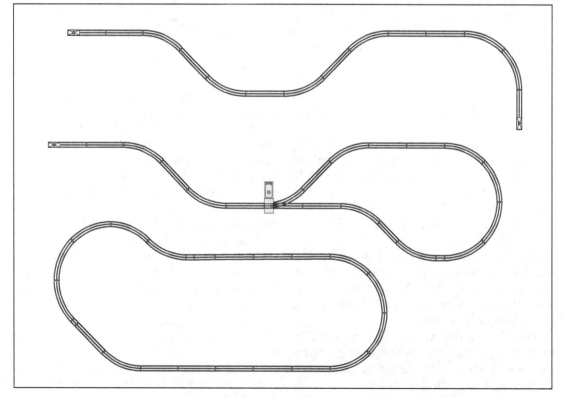

Above: Here is an example of a yard using toy train track.

There are three types of track plans: point to point (top), out and back (middle), and continuous run (bottom).

Bridge out!

Bridge out! Those two short words mean big trouble. If you are driving a car, it means a winding, backtracking detour to get to the other side of the bridge. If you're a locomotive engineer, it means a winding, backtracking detour and double the traffic. Let's not even mention a detour without sufficient passing sidings to handle the traffic increase.

Without bridges there wouldn't be a network of railroads, and our O gauge pikes wouldn't be nearly as exciting. Here a BNSF diesel leads a freight train across the St. Croix River at Hastings, Minn. *Tom Danneman*

This track plan, which fits smartly within a 12- by 12-foot space, depicts a double mainline minus a key bridge. All trains must pass through crossover switches to squeeze through a single Lionel bascule bridge while a new bridge across the river is under construction.

In addition to the bascule bridge—which literally divides the layout into two sections—there is an oil town at one end of the layout and a rural area at the other. While the town isn't quite as big as Houston, it's got a Texas-sized quantity of oil-themed accessories—pumps, derricks, tanks, an oil drum loader, and even an MTH operating gasoline station.

So even as the pump price at the Mobil station down the street climbs above $2 a gallon, you can take comfort in the fact that your O gauge oil empire can still produce hi-test sold at the MTH station for 32 cents a gallon.

Track and wiring

This track plan, as shown, uses traditional tubular track with Lionel and K-Line O-72 and O-42 track switches. It can be built with any brand of track that offers O-42, O-54, and O-72 curves and the aforementioned track switches.

You won't need to mix and match curve sizes here. The loops at either end of the layout are O-42, the gentle curves connecting the loops are O-72, and the only O-54 sections of track are at the approach to the missing bridge on the far side of town.

The back-to-back O-72 track switches on both sides of the river are positioned for high-speed running. Look closely and you'll see that there are no "S" curves that could lead to derailments.

There's plenty of flexibility in this track plan at the bridge. While I've selected a modern-era reproduction of Lionel's operating bascule bridge, you can use a Lionel lift bridge, Lionel or K-Line swing bridge, or even one of MTH or Lionel's Hell Gate bridges. You also can swap bridge locations, moving the existing bridge to the back and the missing bridge to the front. If you do that, you'll need to reverse the two sets of crossover switches. Also note that

there's half of a third rail line across the river. It's used for the construction train and the river barge.

There are no grades to contend with on this track plan; all of the track as shown is on an even plane.

Conventional-control wiring calls for blocks at the missing bridge, the tracks behind the passenger station, and the oil-refinery line. You'd also be smart to block off the mainline into three sections to allow you to hold a train in the passenger-station loop or the rural loop at the far end of the layout. The track plan uses a pair of MTH Pennsylvania Railroad-style signal bridges at the approaches to the bridge. There are plenty of places to add more.

Freight and passenger trains

There's big oil-refinery action and a big dilemma at the river crossing on this track

ACCESSORIES

QTY.	DESCRIPTION
1	Lionel 115 station
1	Lionel 128 newsstand
2	Lionel 156 freight platform
1	Lionel 9220 milk car platfrom
1	Lionel 12948 operating bascule bridge
1	Lionel 2324 operating switch tower
1	Lionel 12847 icing station
4	Lionel 12953 tall Linex oil tank
2	Lionel 12954 wide Linex oil tank
1	Lionel 14155 395 floodlight
1	Lionel 14154 193 water tower
2	Lionel 12902 Marathon oil derrick
1	Lionel 12912 pumping station
1	Lionel 2300 oil drum loader
7	Lionel 12927 yard light
1	Plasticville 1500 diner
1	Gilbert 583 magnet crane
1	MTH 30-9116 oil storage tank
1	MTH 30-9117 storage tank station
1	MTH 30-9096 CJ's textile factory
2	MTH 30-11030 PRR signal bridge
1	MTH 30-90032 Fairview Depot
1	MTH 30-50002 16-inch fence
1	MTH 30-9106 Esso service station
1	MTH 30-9018 hardware store

Modern bridge construction on the Canadian Pacific. Note the barges, cranes, girders, and other pieces of equipment.
Tom Danneman

plan. That calls for a big location—Texas. In days gone by, Texas was served by the Santa Fe, Katy (Missouri-Kansas-Texas), Texas & Pacific, Missouri Pacific, Frisco, Southern Pacific, Burlington Route, and other roads. Today, that translates primarily into BNSF and Union Pacific.

Since this is oil country, O gauge tank cars from postwar Lionel to contemporary K-Line will be right at home on this track plan. With a modest amount of work, you can even match the oil-related accessories to your rolling stock (Sunoco, Gulf, or how about the fictional "JeTexas" oil company from the Rock Hudson/James Dean screen epic *Giant*?)

Beyond bridges and oil, there's a first-class passenger station on this track plan, which calls for topflight passenger service. The tomato red *Texas Special* streamliner is a natural, as are sleek trains bearing the names Santa Fe, Southern Pacific (in orange *Daylight* garb), and Missouri Pacific (in handsome blue).

The design of the oil accessories, station, and bridge allow plenty of era flexibility. Steam, transition-era, or contemporary locomotives fit this track plan. Don't forget to add a pair of water towers for any thirsty steamers.

Scenery and accessories

Speaking of thirsty, it can get pretty dry in parts of Texas. The river as depicted on the track plan has a healthy flow, but you can model a dry riverbed with ease—although you'll need to skip the equipment barge.

If you have a hankering for model cranes, such as those sold by model construction-equipment makers such as Norscot (see the ads in *Classic Toy Trains* magazine), this layout is for you. Next to the barge on the track plan is an American Flyer fixed-base magnetic crane. You can replace it with any number of fixed-base or mobile cranes. Most bridge-construction

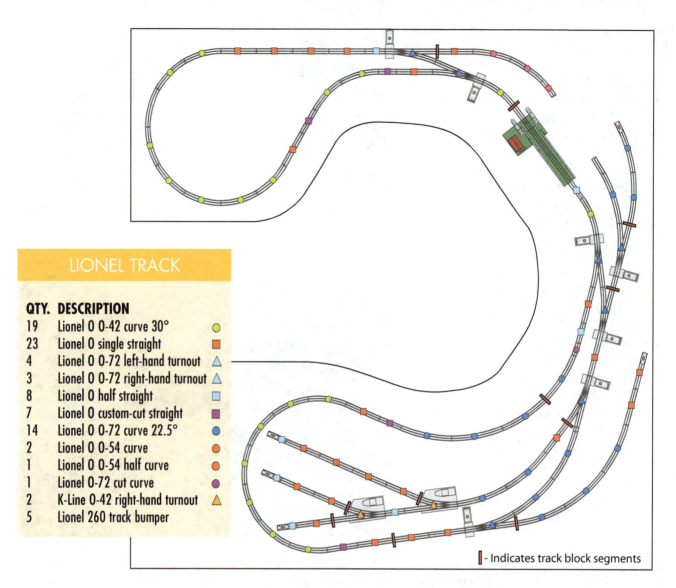

LIONEL TRACK

QTY.	DESCRIPTION	
19	Lionel O 0-42 curve 30°	🟢
23	Lionel O single straight	🟧
4	Lionel O 0-72 left-hand turnout	🔼
3	Lionel O 0-72 right-hand turnout	🔼
8	Lionel O half straight	🔷
7	Lionel O custom-cut straight	🟪
14	Lionel O 0-72 curve 22.5°	🔵
2	Lionel O 0-54 curve	🟠
1	Lionel O 0-54 half curve	🟠
1	Lionel 0-72 cut curve	🟣
2	K-Line 0-42 right-hand turnout	🔼
5	Lionel 260 track bumper	

▯▮ - Indicates track block segments

sites feature multiple cranes, including one on the barge itself. For night work, I've added four modern-era Lionel yard lights to the construction scene.

The big town at one end of the layout uses a Lionel no. 115 city station (or a modern reproduction from MTH, Lionel, or T-Reproductions). Next to the station is a Lionel animated newsstand with that crazy revolving dog and a pair of Lionel station platforms. Beyond the newsstand is everyone's favorite Plasticville diner. Keep in mind that K-Line's gleaming Starlite diner, with its animated red roof sign, would look fantastic in this spot, too.

Behind the station you'll find more Lionel accessories. These include an icing platform (it gets hot in Texas!), milk-car platform, floodlight tower, and industrial water tower.

To give the sense of a city beyond the passenger station, there's a short stretch of highway lined with MTH RailKing buildings and an MTH operation gas station.

At the refinery I recommend installing two operating oil derricks, an oil-pumping station (either Lionel originals or MTH and Lionel reproductions), six Lionel Linex oil tanks (four tall tanks and two wide tanks), an MTH operating oil storage tank, and an MTH storage tank station.

At the city-side approach to the bridge is an operating Lionel switch tower, and all by its lonesome self at the rural end of the layout is an MTH RailKing Fairview Depot.

Yet even with all of these accessories, there is room for a bit of greenery on this walk-in layout. On both riverbanks are groves of trees. There are more trees within the rural return loop, lending the sense that this railroad really does leave town.

Phew! This is a lot of layout. And to think, it was all inspired by just two words: Bridge out!

Cold Mountain

If you've been a flatlander for most of your life, the mountains always beckon. If you live in the mountains, well, you're already there. This O gauge track plan, inspired by today's Burlington Northern Santa Fe Railroad in the Dakotas and Montana, features a run from the prairie through the mountains and back.

Designed with MTH RealTrax, this winding figure-eight layout appears simple, but its scenery—a mountain range with twin tunnels cutting diagonally across its middle—creates two distinct areas. To the west of the twin central tunnels are a crossover bridge and the rugged Rocky Mountains. There are no towns here, but a fast-flowing mountain stream—justification for a pair of bridges—and another quick tunnel add plenty of interest. They also give your scenery skills a welcome workout.

Back across the divide, on the east side of the layout, is the remainder of the figure-eight, stretched into a kidney shape on a rolling prairie. A long passing siding, ending just east of one of the tunnel portals, allows two full trains to be on the layout at the same time.

On the edge of town is a small yard or industrial area, with a lead track requiring a reverse movement. If you have the space, the yard can be lengthened or widened to accommodate more track or additional accessories.

A contemporary Burlington Northern Santa Fe diesel sweeps across a valley pulling a string of auto-loader cars through the northern Rockies. *Tom Danneman*

How's this as a reference point for an up-and-over crossing? Two Union Pacific trains cross paths on former Missouri Pacific and Missouri-Kansas-Texas tracks near New Braunfels, Tex., in 1990. *Carl M. Lehman*

ACCESSORIES

QTY.	DESCRIPTION
1	Lionel 464 operating sawmill
1	Lionel 3656 operating stockyard
1	MTH 30-9023 row house 1 (white)
1	MTH 30-9024 row house 2 (gray with porch)
2	MTH 30-9044 row house 1 (brown)
1	MTH 30-9045 row house 2 (cream with porch)
1	MTH 30-9076 row house 2 (red)
1	MTH 30-9089 row house 2 (green)
1	MTH 30-9129 operating watchman's shanty
1	MTH 30-90004 country freight station
1	MTH 30-90008 workhouse 1 (yellow and red)
1	MTH 30-90013 yard office
1	MTH 30-90026 granary
1	MTH 30-90032 Fairview depot
2	MTH 40-1014 girder bridge (10 inches)

As shown, this irregularly shaped walka-round track plan, at its broadest points, is 12 feet wide and 12 feet long.

Track and wiring

The track plan is designed for either clockwise or counterclockwise train movement. Depending on your tastes, however, you may wish to move or even reverse the track switch that leads to the yard.

If you follow my design, you may find it handy to base an old switching locomotive in town to handle forward-and-reverse movements through the yard.

Another thought, if space allows, would be to add a turntable and roundhouse. A locomotive facility amid the mountains calls to mind Union Pacific shops in Cheyenne, Wyo., and other destinations along the railroad's transcontinental route.

While the track plan shown here utilizes MTH's snap-together RealTrax, almost any other brand of track can be substituted. Excluding the siding, all curves are O-54 or greater and most feature O-72 easement sections. All track switches are O-72 for derailment-free operation.

Note the grades needed for the overpass on the far side of the mountain ridge. A train running clockwise on the far loop begins to climb just after ducking below the overpass bridge. The train makes a 270-degree turn as it climbs across three short bridges and into a tunnel portal. The downgrade actually begins within this tunnel, gently descending to the base level of the layout at the town's station, just before the entrance to the yard.

Block wiring is straightforward. Both the passing siding and adjacent main line are separate blocks. There's another block on the rural end of the layout beyond the stream—just in case traffic backs up in town and an arriving locomotive needs to cool its heels for a while.

In the yard, the approach area is a separate block, as are each of the four tracks. This arrangement facilitates switching movements while other trains are operating on the layout.

Freight and passenger trains

If you're not a BNSF or UP fan, keep in mind that fallen-flag roads—the Northern Pacific, Great Northern, Burlington Northern, Denver & Rio Grande Western, and even the Santa Fe—crossed the Rocky, Cascade, and Sierra Nevada ranges. Or you can locate this layout north of the border, with service from the Canadian National and the Canadian Pacific. There are also regional lines in the mountains, like Montana RailLink.

The Milwaukee Road, among the last of the transcontinental routes, electrified part of its line to the west. It's easy to envision MTH RailKing catenary and big electric

This isn't just any ordinary tunnel running through the northern Rockies. Here, a Canadian Pacific grain train crosses beneath itself at the exit to the railroad's famed lower spiral tunnel in 1986. *Alex Mayes*

Heading west in the summer of 2004 toward the northern Rockies and onward to Seattle, Wash., is the Amtrak *Empire Builder*. *Cody Grivno*

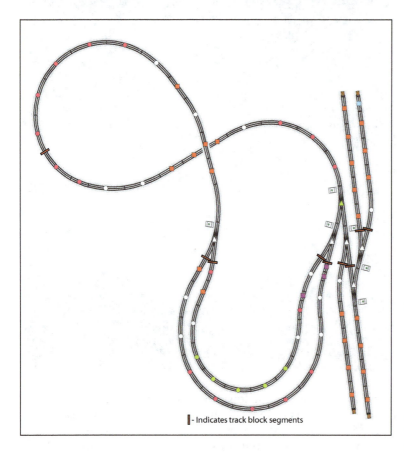

- Indicates track block segments

MTH TRACK

QTY.	DESCRIPTION	
24	MTH 40-1001 straight (10 inches)	🟠
14	MTH 40-1010 O-72 curve	⚪
1	MTH 40-1012 straight (5.5 inches)	🟦
3	MTH 40-1018 straight (3.5 inches)	🟪
5	MTH 40-1020 O-72 right-hand switch	△
1	MTH 40-1021 O-72 left-hand switch	△
4	MTH 40-1024 track bumper	
5	MTH 40-1042 O-42 curve	🟢
15	MTH 40-1054 O-54 curve	🟠
3	MTH 40-1057 O-54 half-curve	🟠

locomotives blasting through tunnels and across cold mountain streams.

Freight loads can include grain hoppers from the prairie, along with coal and timber products. Then you can have automobiles heading west from manufacturing plants in Detroit.

But all is not freight on this track plan. Today, Amtrak's Empire Builder, with dual-level passenger cars, runs through this territory. In years past, trains such as Northern Pacific two-tone green streamliners, the Great Northern's original green-and-orange Empire Builder, and the Milwaukee Road's famed Olympian Hiawatha, with its classy

glass observation car, passed through the plains and mountains.

Scenery and accessories

Each of the three tunnels is short—the longest just three lengths of track—so if by chance there's a derailment in a tunnel, you'll easily be able to fish out cars and locomotives. Tunnel liners are a must, since daylight will brighten the tunnel interiors and show the inside of your plaster or pink-foam mountains.

The mountain ridge should be tall. Think big, then add another foot. You want to visually justify the ridge's two tunnels, since your railroad's engineers didn't specify deep cuts open to the sky instead. Rocks (perhaps a rock slide?) and fir trees round out the picture. The mountain stream should be narrow and deep to suggest swift water. A scene with hikers or canoes would be a plus.

The town, as shown, is populated with RailKing, Lionel, and other brands of buildings, all readily available at hobby shops and other toy train outlets. Its streets call for a pair of rail/highway crossings, and the town can accommodate an operating accessory or two, such as an MTH firehouse or a gas station. Operating accessories shown here include a Lionel cattle loader and operating sawmill.

There's a temptation to add more track to this plan—after all, the layout as designed will fill a significant portion of a 12- by 12-foot room. But don't—sometimes less is more. With a clever rocky ridge dividing the layout into two distinct areas, there's plenty here to satisfy both flat-landers and mountaineers.

Small Town

In the 1980s, rock singer John Mellencamp wrote an ode to small towns throughout America entitled, appropriately enough, "Small Town." The lyrics go something like this: "I was born in a small town ... and I live in a small town ... prob'ly die in a small town ... and that's good enough for me."

To the best of my knowledge, Mellencamp doesn't have an O gauge layout, but if he did, the Indiana native's layout would *prob'ly* look something like this.

This isn't a small town in Indiana, but it's the next best thing: a small town in Ohio. A Baltimore & Ohio 2-10-2 crosses Park Street in Wapakoneta, Ohio, in the mid-1950s. Note that the crossing signals (dead ringers for the postwar Lionel no. 155 ringing signals) are placed in the center of the street. *Linn Westcott*

This seemingly simple 9- by 12-foot walk-in track plan includes four crossovers, two reversing loops, and, as depicted here, a small town.

Track and wiring

This track plan is best described as a bent dog-bone. It's designed for running, and there are only two lines branching off a single spur track on the left lobe of the layout. Trains can run clockwise, counterclockwise, and back again, thanks to two reversing loops.

Where are those reversing loops? Hidden within the swooping lines that cross over each other in three places in the middle section of the track plan. Trace your finger around the track plan and you'll discover that the two return loops even share some of the same rails. It's a good track plan if you want to keep your friends guessing.

Designed using regular Lionel O gauge track, this plan calls for five O-72 track switches and one O-31 track switch. It's one of the most economical plans in this book, which is good because it leaves you plenty of cash to build a downtown at one end of the layout and an industrial area on the other.

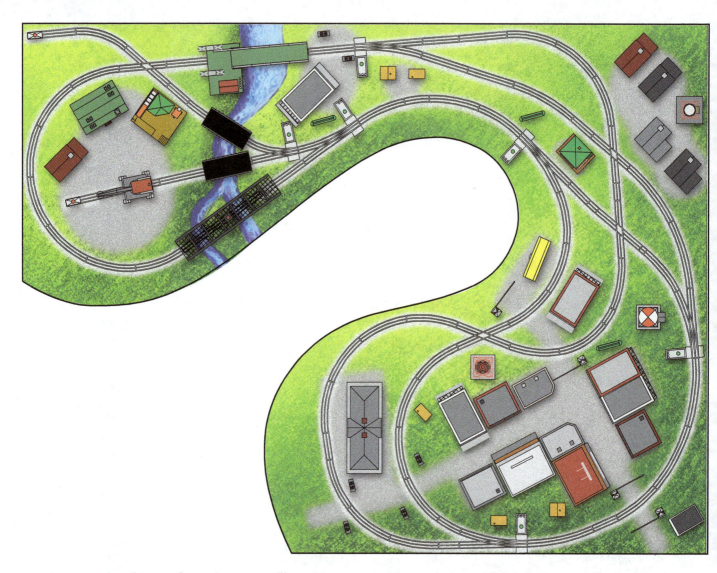

To get everything to fit as shown, you'll need to cut seven straight sections and three curved sections— an O-72, an O-42, and an O-31—to non-standard lengths. All the other curves are O-42 or wider, but there are two pieces of O-31 track that connect two of the 45-degree crossings on the edge of downtown.

Conventional-control wiring calls for separate blocks for each of the lines at the industrial end of the track plan and the two sections of open track in front of and behind the town's passenger station. These sections can be used to hold trains while moving other locomotives on the layout.

You can get away with a mid-sized transformer for this layout. All the same, it's always nice to have the extra power output of a new or old Lionel ZW, an MTH Z-4000, or the new MRC Pure Power Dual.

Freight and passenger trains

John Mellencamp is from Indiana, so that's where this fictional O gauge town—I'll call it Middleburg— is located.

A wealth of railroads ran through the state of Indiana. Big boys like the New York Central and the Pennsylvania Railroad passed through on the way to Chicago. Indiana was a home to the beloved Monon as well as the Erie, Chesapeake & Ohio, Nickel Plate Road, and the Baltimore & Ohio.

And who can forget the Wabash? The term "fallen-flags"—coined to describe railroads no longer in existence—originated with the demise of the Wabash and its waving flag emblem. Heck, even the Southern Railway had a line through the lower part of Indiana.

With all of those railroads and their contemporaries—Conrail, CSX, and Norfolk Southern—any freight train short of ore cars will be appropriate for the rails of Middleburg. Given the tight clearances through town, sticking with Geeps, F units, and other four-axle diesels—or similar-sized steamers— makes the most sense.

Middleburg being a small town, passenger service should be appropriately modest. Rail Diesel Cars, doodlebugs (MTH makes a dandy one), and two- or three-coach trains (just like in the old Lionel catalogs) will work well snaking their way through the crossings and return loops.

Conrail freight trains ran right down the middle of the street in the small town of West Brownsville, Pa. *D. Jacobson*

ACCESSORIES

QTY.	DESCRIPTION
1	Gilbert 772 water tower
1	Lionel 193 industrial water tower
3	Lionel 310 billboard set
1	Lionel 2300 oil drum loader
1	Lionel 2324 operating switch tower
1	Lionel 12772 extension truss bridge
1	Lionel 12834 PRR gantry crane
1	Lionel 12948 operating bascule bridge
2	Lionel 22907 die-cast girder bridge
1	MTH 30-9012 corner drugstore
1	MTH 30-9018 hardware store
1	MTH 30-9051 Myerstown station
1	MTH 30-9078 townhouse 1
1	MTH 30-9082 hotel
3	MTH 30-9096 CJ's Textiles factory

QTY.	DESCRIPTION
2	MTH 30-90008 workhouse 1 (yellow and red)
1	MTH 30-90009 workhouse 1 (blue and black)
1	MTH 30-90010 workhouse 2 (gray and black)
1	MTH 30-90011 workhouse 2 (cream and red)
2	MTH 30-90017 Feather & Son city factory
1	MTH 30-90019 Lombardi's Pizza
1	MTH 30-90020 Katz's Deli
1	MTH 30-90023 soda fountain corner
1	MTH 30-90026 granary
1	MTH 30-90034 McHale's Tavern building
4	MTH 30-11012 operating crossing gate signal
6	MTH 30-11014 operating crossing flasher set
1	MTH 30-11051 Flyer 23772 water tower
1	Plasticville 1500 diner
5	Plasticville 1627 hobo shack

Scenery and accessories

So far, all of the references to small towns in this chapter have been words and song. Only after you add scenery and structures to this track plan does it take on a life of its own.

Start with downtown Middleburg. The structures depicted are from MTH's RailKing line. At the intersection are MTH's corner drugstore and its soda fountain building. Down Main Street and along both sides of Railroad Avenue are 10 other RailKing buildings, from McHale's Tavern to Lombardi's Pizza to Katz's Deli. Seems all they do in Middleburg is eat and drink.

MTH's Myerstown passenger station, a recent addition to the RailKing line, tucks in nicely between the two tracks looping around the lower end of downtown. MTH signals and gates protect the road crossings on Main Street and Railroad Avenue.

While not depicted on the track plan, downtown also calls for plenty of streetlights. In fact, between the illuminated structures and streetlights, you'll be wise to use an auxiliary transformer to power all the accessories.

Between downtown and the industrial area are plenty more items. There's a Lionel operating switch tower, an American Flyer water tower, Lionel billboards, two Plasticville hobo shacks, a row of MTH

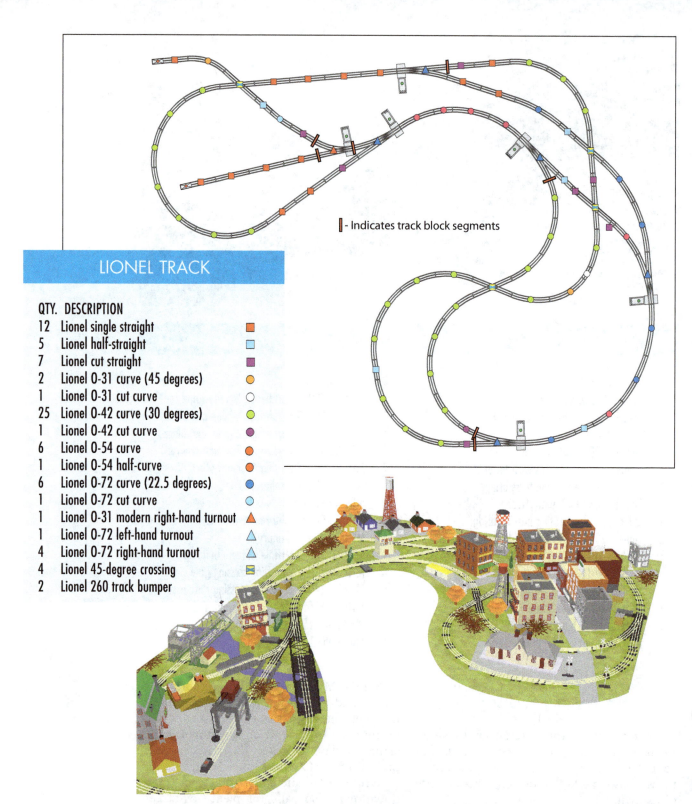

- Indicates track block segments

LIONEL TRACK

QTY.	DESCRIPTION	
12	Lionel single straight	■
5	Lionel half-straight	◻
7	Lionel cut straight	◼
2	Lionel 0-31 curve (45 degrees)	●
1	Lionel 0-31 cut curve	○
25	Lionel 0-42 curve (30 degrees)	●
1	Lionel 0-42 cut curve	●
6	Lionel 0-54 curve	●
1	Lionel 0-54 half-curve	●
6	Lionel 0-72 curve (22.5 degrees)	●
1	Lionel 0-72 cut curve	○
1	Lionel 0-31 modern right-hand turnout	▲
1	Lionel 0-72 left-hand turnout	△
4	Lionel 0-72 right-hand turnout	△
4	Lionel 45-degree crossing	⬓
2	Lionel 260 track bumper	

RailKing workhouses, and my favorite, a Plasticville diner.

The industrial end of the layout features a river—the only challenging scenic feature beyond the autumn and winter trees depicted on the track plan. Four Lionel modern-era bridges cross the river: a truss bridge, a bascule bridge, and for the spur, a pair of girder bridges. The spur features a Lionel gantry crane, an operating oil drum loader, and two MTH factory buildings.

Beyond trees and the river, you'll need to make some city streets, sidewalks, dusty lots, and plenty of junk metal for the scrap yard and river area.

To paraphrase our singer Mellencamp, "Now go have yourself a ball with this small town"—track plan, that is!

Old King Coal

For as long as there have been toy trains, there has been an equal fascination with coal. In the prewar and postwar years, Lionel produced no fewer than six coal loading or unloading accessories and a half-dozen or more coal dumping freight cars. Two of American Flyer's tallest and most endearing accessories were the nos. 785 coal loader and towering 752 Seaboard coaler.

Coal mining was a risky business. Here a Chessie System train with a load of empty hoppers swings around "graveyard curve" near Terra Alta, W. Va., in the early 1980s. Scenic Express, a model railroad scenery supplier, sells O scale headstones that allow you to duplicate this scene. *Mark Perri*

Lionel has reissued its nos. 397 diesel-type coal loader, 456 coal ramp, and 497 coaling station. MTH has reproduced the Flyer coal loaders. K-Line has developed its own modern interpretation of a coal ramp, and Lionel has produced a variation of its gantry crane that replaces the crane's magnet and hook with a clamshell bucket—to scoop up coal.

For those of you who share this fascination with coal—and there are plenty of you—take a gander at this 9- by 12-foot walkaround track plan, which uses four of Lionel's best coal-handling accessories. Beyond coal, this plan offers three different routes, dual reversing loops, a 3½-foot-long central bridge above a crossing and a coal spur, a 10-foot-long passing siding, and plenty of green Appalachian scenery.

Track and wiring

This track plan was designed using Atlas O nickel-silver track. All three lobes use O-45 (that's 45-inch-diameter) curved track. There are five O-45 track switches, a pair of O-36 switches, and one Atlas O no. 5 short left-hand switch. Rounding out the package are 60-degree crossing and 45-degree crossing sections.

To get everything to fit, especially adja-

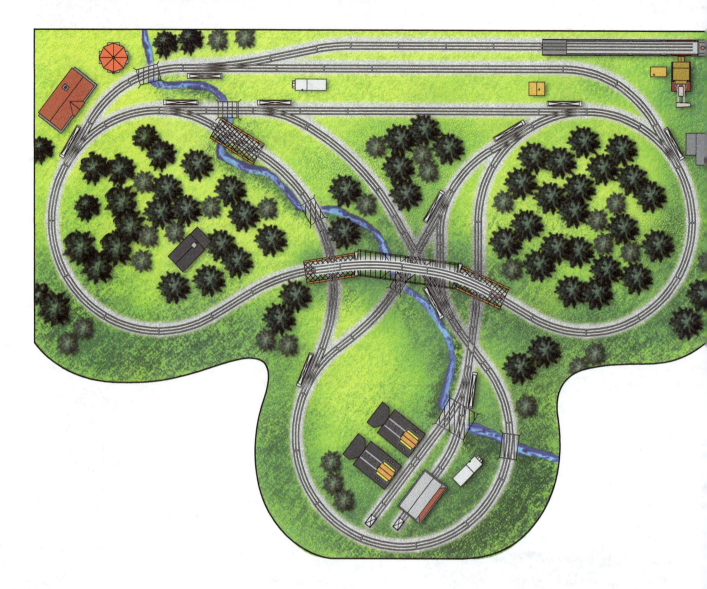

cent to the 45-degree crossing track beneath the main bridge, you need five curved sections cut to non-standard lengths and several straight sections cut to fit. Atlas O didn't design its track to be cut, but its track joiners—they slide onto the ends of the rails—readily accommodate cut rails.

It doesn't matter that you'll be cutting off the male/female snap-together sections of the plastic track ties. On a permanent layout, the track will be affixed to the roadbed anyway.

You can build this layout with another brand of track. However, because Atlas O's curve sizes (36, 45, and 63 inches) are atypical, using something else means you'll have to modify the plan rather than building a carbon copy of it.

There are mainline grades on both sides of the big central bridge. To keep the climb reasonable, you should put as many sections of track as possible on a grade.

On this track plan, the two track switches on the back corners of the layout, and even the first and last sections of the passing siding, are part of the ascending and descending grades. Of course, steep, winding curves are what coal country railroading is all about.

Wiring this track plan for conventional-control operation calls for several independent blocks. These include the central loop, the crossing within the loop, the larger of the two approaches to the loop, the passing siding, the section of the main line adjacent to the passing siding, the coal-ramp spur, and both forks of the coal-loading spur within the central loop.

That's a lot of wire, but doing this lets you alternate between two mainline trains, one with a home base on the passing siding and the other with a home base on the central loop or the left-side approach to the loop. This sounds more confusing than it is,

ACCESSORIES

QTY.	DESCRIPTION
3	Gilbert 571 truss bridge
2	Lionel 96/97 coal elevator
1	Lionel 456 coal ramp
1	Lionel 2175 gravel loader kit
1	Lionel 2315 coaling station
1	Lionel 12734 passenger/freight station
1	Lionel 12916 "138" water tower
2	MTH 30-90005 stainless mobile home
1	MTH 30-90009 workhouse 1 (blue and black)
1	MTH 30-90010 workhouse 2 (gray and black)
2	Plasticville 1627 hobo shack

but the more track blocks you create, the more operation possibilities you create.

Freight and passenger trains

Coal, naturally, is king on this layout. MTH, K-Line, and Lionel make operating dump cars, and several styles of dump cars were made during the postwar era. As far as pure hoppers go, K-Line and MTH make rock-solid die-cast metal hoppers, and Lionel, MTH, K-Line, Williams, and Weaver make plastic-bodied hopper cars in a variety of styles and dozens of road names.

I designed this layout with the kingpins of eastern coal in mind. Since this track plan has plenty of space for two trains to coexist on the rails, it creates a natural coal-hauling rivalry. Think the Baltimore & Ohio versus the Chesapeake & Ohio, or the Virginian versus the Norfolk & Western.

We can't forget coal-hauling locomotives. Lionel, MTH, and K-Line offer scaled-down O gauge articulated locomotives in eastern coal road names that will be just right pulling strings of hoppers through this layout.

Not into steam? Try lashing up Electro-Motive Geeps or aging Alcos. Into modern power? MTH, Lionel, and Atlas O offer models of recent-vintage diesels from Electro-Motive and General Electric in contemporary blue-and-yellow CSX and black-and-white Norfolk Southern road names.

Passenger service can be as bold or subtle as you like. Bold, as in yellow-and-blue Chesapeake & Ohio streamliners, or subtle like the deep maroon of the Norfolk & Western or the royal blue and white of the Baltimore & Ohio.

Scenery and accessories

Coal country means steep hills and plenty of greenery. The centers of the loops climbing and descending from the central bridge are steeply sloped and filled with deciduous trees, with a handful of pines thrown in for variety.

While this track plan depicts spring, such a layout will look dynamite in the yellows, oranges, and reds of autumn. In the central loop are exposed rocks and boulders, and if you use a pair of Lionel no. 96 coal loaders (as shown on the track plan), you'll want to make this area look like the entrance to a mine.

A small creek runs through middle of the track plan, giving justification for several small bridges or scratchbuilt trestles. One bridge crossing the stream is a no. 571 American Flyer truss bridge (there are two more no. 571s on the approaches to the large central bridge).

S gauge American Flyer bridges with O gauge track? Yes, the 571 bridge is plenty wide for O gauge rolling stock; molded in

Top: CSX locomotives dig in while rounding a curve near Chapmanville, W. Va. Note the short utility poles to the right.
Everett N. Young

Bottom: Why are there several Plasticville hobo shanties on this track plan? Because they resemble real-life railroad structures, like this switch tender's shanty in Coxton, Pa. Did you see the bench and the kitchen chair to the right?
H. Russell

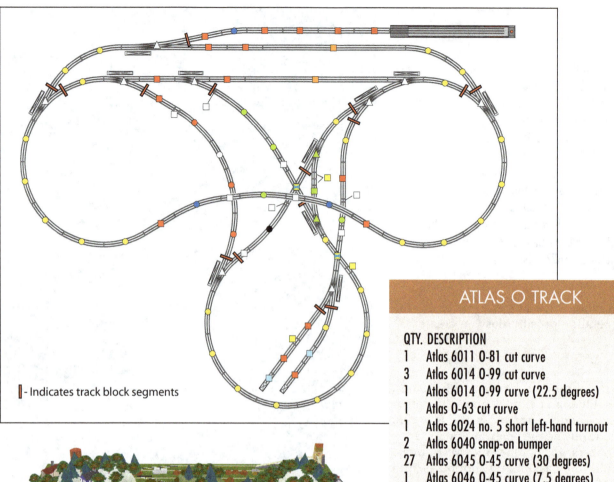

▮ - Indicates track block segments

red or orange, it's a handsome, open-arch alternative to the garden-variety plate girder bridge. Lionel never made anything like the 571.

Speaking of bridges, the main span above the crossing is on a subtle curve, so not just any out-of-the-box bridge will work. Kitbashing or adding a combination of bridges and trestles may be in order to keep the sides of trains from scraping the vertical members of the bridge.

On the backside of the layout are more trees, big boulders, and a lone MTH RailKing workhouse on one of the hills. Along the main line is a modern-era Lionel freight/passenger station and an adjacent operating water tower.

At the other end of the layout is the raised coal ramp and a sprinkling of Plasticville hobo shacks (can't have too many of those!). If you have more space, the area around the coal ramp is ripe for additional structures and accessories.

But don't get too carried away with structures on this layout. Otherwise, you'll lose the sensation that coal is king in this part of the country.

Big City

Badda-bing, badda-boom! This here's your mainstream North Jersey/Long Island O gauge track plan. Two trains runnin', plenty of those whatcha-ma-call-its—yeah, accessories—and all in a 6- by 12-foot space, with an additional 3- by 4-foot industrial yard.

How's this for big city railroading? The morning rush hour winds down at the Long Island Railroad yard in Long Island City in June 1986. In the background is the Manhattan skyline and to the left are cars many of us once owned and may wish to forget—Thunderbirds, Granadas, and Chevettes. *George A. Forero Jr.*

Yo, this thing looks like the Long Island Railroad taking suits into the city, or NJ Transit zippin' through places like Secaucus, Maplewood (yeah, that was the name of a Lionel passenger car, too), and Newark. Speakin' of Newark, the only thing this track plan ain't got is an airport!

Whoa! Let me slow down. Please excuse the vernacular, but the native New Jersey in me got me all juiced up drawing this track plan. It's got everything:

- A single level with no grades
- Two trains on inner and outer loops
- A figure-eight central crossing, accessed in one direction from the inner loop and in the other direction from the outer loop
- Two passenger stations with plenty of additional platforms
- Three industrial areas
- And more operating accessories than you can shake a stick at.

Track and wiring

This track plan uses MTH RealTrax, taking advantage of its O-42 and O-54 curves. The outer loop is all O-54, so it can handle any locomotives and freight cars other than the big O-72 boys. Inside, the smaller loop is all O-42, including track switches.

The central crossing is comprised of a pair of 45-degree crossing sections and one

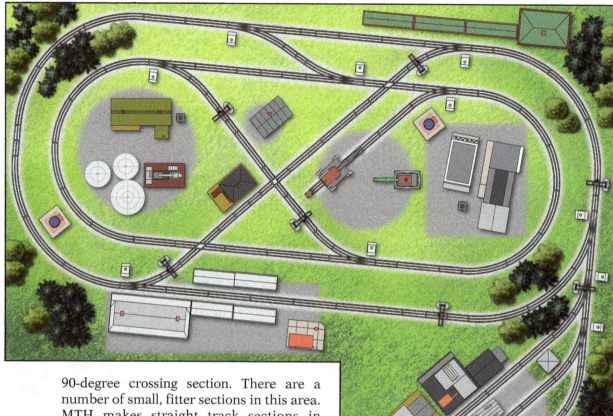

90-degree crossing section. There are a number of small, fitter sections in this area. MTH makes straight track sections in 3.5-, 4.25-, 5-, and 5.5-inch lengths, and this layout uses them all. Six O-42 and one O-54 half-curved sections, also in MTH's catalog, round out the atypical track pieces used to build this layout.

Within the inner loop is a short spur slanting off the only O-31 track switch on the layout. The track leads to a scrap yard with a Lionel gantry crane.

A three-track industrial yard occupies a 3- by 4-foot outrigger section to the right of the main layout.

Wiring this plan for conventional-control operation is a busy endeavor, but not overwhelming. For two-train operation, the inner and outer loops should be independent blocks. Each loop is further divided into thirds so trains can switch places on the inner and outer loops.

As for the industrial yard, all three tracks and the approach track are separate blocks. While all of these blocks may seem like overkill, you'll thank yourself later when operating trains.

Freight and passenger trains

The track plan as designed is set up for busy commuter service with two stations and seven platforms. Local passenger trains can keep to the outer loop and the two sta-

tions, while express trains can stay on the inner loop. Running one train clockwise and the other counterclockwise can simulate morning or evening rush hours.

As for locomotives and rolling stock, there's plenty to choose from. In the late 1990s, Lionel produced several uncataloged NJ Transit sets, and K-Line offered a Long Island set. Atlas O and MTH have marketed commuter equipment, too.

And who says this layout must be set in the present? There have been commuter trains for as long as there have been railroads. The Pennsylvania, Lackawanna, Jersey Central, New York Central, Long Island, and New Haven all ran rush-hour trains in and out of the Big Apple.

And those somber green or Tuscan-colored coaches weren't just pulled by diesel units. The Pennsy used its famous K4 steam locomotives in commuter service. The Jersey Central assigned Camelbacks to this task. Or, picture this layout with MTH's overhead catenary and Pennsylvania GG1 or New Haven EP-5 "jet" electric locomotives. Badda-bing!

This time it's Jersey City—not New York—in the background as a Jersey Central steam locomotive works the coach yards in the mid-1940s. Just visible to the left of the locomotive is a Camelback steamer. Also note the Lionel no. 195-style floodlight towers in the yard. *William R. Frutchey*

While passenger operation is the main focus, there's plenty of room for mixed freight trains dodging between the two loops to pick up gondolas at the central scrap yard. In the industrial yard, you can spot hoppers at the gravel tower or boxcars at the operating forklift platform.

Scenery and accessories

This track plan is big on operating accessories. The scrap yard relies on a modern-era Lionel gantry crane and an American Flyer no. 583 fixed-base crane. The industrial sidings feature a modern-era Lionel gravel loader and a reissue of a postwar no. 264 automatic forklift platform.

On the northeast end of the outer loop are a Lionel no. 132 station and three no. 157 platforms. On the opposite side of the layout are an MTH two-story brick station and four matching platforms, complete with an postwar animated newsstand (you still gotta love that spinning dog!) and an operating Lionel station clock.

Inside the loops, in addition to the two cranes at the scrap yard, are two operating accessories that don't need rail sidings: a modern-era Lionel oil pumping station and an MTH RailKing operating transfer dock. Also in the central area are two blinking water towers and a pair of the new Lionel operating industrial smokestacks.

If that's not enough, the operating accessories are surrounded by non-operating structures: Lionel Linex fuel tanks next to the oil pump, an MTH factory next to the transfer dock, and an MTH brewery next to the forklift platform.

ACCESSORIES

QTY.	DESCRIPTION
1	Gilbert 583 electromagnetic crane
1	Lionel 128 animated newsstand
1	Lionel 132 passenger station
3	Lionel 157 station platform
7	Lionel 452 gantry signal
1	Lionel 2175 gravel loader kit
1	Lionel 12773 freight platform kit
1	Lionel 12834 PRR gantry crane
1	Lionel 12901 steam shovel kit
1	Lionel 12905 factory kit
1	Lionel 12912 oil pumping station
2	Lionel 12953 tall Linex oil tank
1	Lionel 12954 wide Linex oil tank
1	Lionel 14000 "264" operating forklift platform
2	Lionel 14142 operating industrial smokestack
1	Lionel 14147 station clock
2	Lionel 14154 "193" industrial water tower
1	Lionel 22902 Quonset hut
4	MTH 30-9006 passenger station platform
1	MTH 30-9014 passenger station with platforms
1	MTH 30-9031 PRR switch tower
1	MTH 30-9084 brewery
1	MTH 30-9096 CJ's Textiles factory
1	MTH 30-9110 operating transfer dock
1	MTH 30-90004 country freight station
1	MTH 30-90006 trackside yard tower

Trackside, seven postwar Lionel no. 452 gantry signals protect the central crossing and both approaches to the industrial yard.

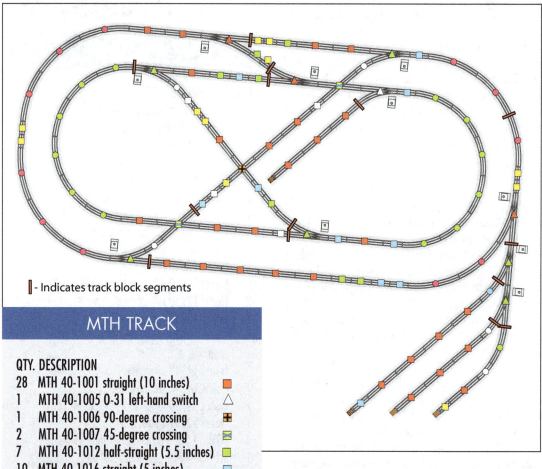

| - Indicates track block segments

MTH TRACK

QTY.	DESCRIPTION	
28	MTH 40-1001 straight (10 inches)	🟧
1	MTH 40-1005 O-31 left-hand switch	△
1	MTH 40-1006 90-degree crossing	✠
2	MTH 40-1007 45-degree crossing	✕
7	MTH 40-1012 half-straight (5.5 inches)	🟩
10	MTH 40-1016 straight (5 inches)	🟦
4	MTH 40-1017 straight (4.25 inches)	⬜
11	MTH 40-1018 straight (3.5 inches)	🟨
4	MTH 40-1024 lighted bumper	
13	MTH 40-1042 O-42 curve	🟢
2	MTH 40-1043 O-42 right-hand switch	🔺
4	MTH 40-1044 O-42 left-hand switch	🔺
6	MTH 40-1045 O-42 half-curve	⚪
11	MTH 40-1054 O-54 curve	🔴
3	MTH 40-1055 O-54 right-hand turnout	🔺
1	MTH 40-1057 O-54 half-curve	🔵

Scenery on this busy layout is easy. It consists of trees in the four corners, a bit of greenery around the stations, track ballast, a few roads, and plenty of gravel and dirt.

Don't forget to fill the scrap yard with rusty pieces of metal and to fence off both the scrap yard and the oil pump and storage tanks. If your layout is against a wall, you should use a city skyline as a backdrop.

Badda-bing, badda-boom. If this track plan's not a winner for you, well, hey, forgettabout you!

Round the Mountain

In the 1980s, I was in northern California on an extended business trip. During a free weekend, four of us took a day trip to Yosemite National Park. Living in plywood-flat Florida at the time, I'll never forget the massive redwoods and the Ansel Adams vistas that I saw from the rim of Yosemite Valley.

Four Denver & Rio Grande Western diesels pull a train through rugged Glenwood Canyon in the heart of the Rockies in 1986. *Thomas R. Schultz*

And I also won't forget our drive out of Yosemite. It was dusk, and from the backseat I found the endless switchbacks on the two-lane highway (it was a shortcut!) simply fascinating. But our driver, a native Floridian, was utterly unnerved.

Which takes us to this track plan and its long, meandering, and rugged reverse loop. A good look at the track plan shows you a railroad version of our descent out of the Sierra Nevada range 20 years ago!

Track and wiring

Set in the Rockies or the Sierras, this track plan is built from Lionel O gauge track and takes up most of a 9- by 12-foot space. Curves on the main oval are O-54 with O-72 track switches. The lower-level reverse loop, nearest to the wider branch of the river, uses O-54 easement curves, tightening to O-42 sections.

The upper-level reverse loop is a doozy, using nearly 50 sections of track! It climbs in a counterclockwise manner along the back and left edges of the track plan and then crosses the lower-level return loop and part of the river on the first of four bridges.

The track curves tighten from O-54, to O-42, and finally to O-31 as the line weaves in and out, passing through three short tunnels, crossing the river and its tributary

ACCESSORIES

QTY.	DESCRIPTION
1	Gilbert 752 Seaboard coaler
1	Gilbert 789 baggage smasher station
1	Lionel 98 coal bunker
1	Lionel 138 water tower
1	Lionel 356 operating freight station
1	Lionel 464 operating sawmill
1	Lionel 497 coaling station

twice, and crossing the lower-level track three times.

At the center of the layout is a spur track off an O-31 switch. The spur bends around the other side of a peak to an ore mine. Things are so tight amid the shear rock walls and rugged terrain that the mine spur needs a bridge at its cliff-side end just to allow an extra or car or two to be filled!

The entire layout calls for five specially cut straight sections of track, one cut O-72 curve section, and one cut O-54 curve section. It also uses four sections of 40-inch extra-long Lionel straight track.

For conventional-control operation, electrical blocks are needed for the main oval, the lower-level reverse loop, and both sidings adjacent to the operating sawmill and loading accessories at the lower left of the track plan. The ore-mine spur needs a block, as does a short siding near the water tower and the train station.

Ideally, the elevated reverse loop will be divided into three blocks. This allows a train to wait on either side of the ore mine while a switching locomotive clears the line by going down to the water-tower spur.

Freight and passenger trains

There isn't much passenger service in this neck of the woods, but the nature of this track plan cries out for a tourist train, like the Durango & Silverton line in the Rockies. That justifies operating just about anything ever built to haul passengers, from 19th-century Lionel or MTH 4-4-0 steamers and matching open-platform coaches to a modern-day Denver & Rio Grande Western ski train (just watch the overhang on the curves!).

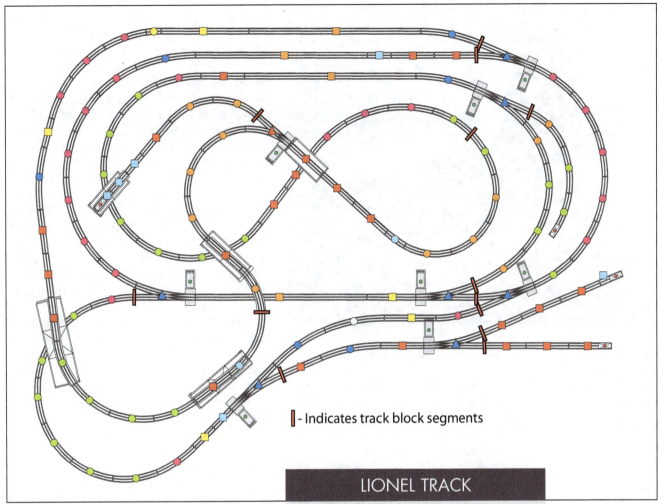

| - Indicates track block segments

LIONEL TRACK

QTY.	DESCRIPTION	
4	Lionel straight (40 inches)	🟧
22	Lionel single straight	🟥
6	Lionel half-straight	🟦
5	Lionel cut straight	🟨
11	Lionel O-31 curve (45 degrees)	🟠
24	Lionel O-42 curve (30 degrees)	🟢
22	Lionel O-54 curve	🔴
2	Lionel O-54 half-curve	🔵
1	Lionel O-54 cut curve	🟡
7	Lionel O-72 curve (22.5 degrees)	🔵
1	Lionel O-72 cut curve	⚪
1	Lionel O-31 modern left-hand turnout	🔺
4	Lionel O-72 right-hand turnout	🔺
3	Lionel O-72 left-hand turnout	🔺
4	Lionel 260 track bumper	

Choose whatever ore you like for the mine—silver, gold, lead, or "toytrainium." While the key accessories on the layout—the prewar Lionel no. 98 coal bunker and postwar no. 497 coaling station, and the American Flyer no. 752 Seaboard coaler—were meant to move coal, they work equally as well with a substance of another color.

The tight curves of the upper reverse loop are perfect for ore cars or postwar-style dump cars. You can readily justify dragging the short cars down the mountain from the mine, unloading them at the coaling station, and then reloading the ore into modern coal hoppers for a trip around the O-54 main line.

Like the ore cars and coal hoppers, locomotives can complement one another. Put a pair of weathered Geeps or a small-drivered steamer in the mountains and some beefy six-axle diesels or one of the semi-scale articulated locomotives from Lionel, MTH, or K-Line on the main line.

O gauge models of alternative steam locomotives, like MTH and Lionel Shays and Heislers—seem perfect for a vertical track plan like this one. But, except for the Heisler, they can't negotiate O-31 curves. K-Line, however, has come to the rescue by releasing a husky two-truck Shay that's designed to handle curves as tight as O-31.

once-mighty stream back in the days of dinosaurs. Steep walls are needed again below the outer edge of ore-mine spur.

Take a trip to your local library and look for books that include photographs of railroads in the Rockies or Sierra Nevadas. After you've seen the photographs, you'll know what I mean by steep.

Fir trees, a river with a tributary stream, and bridges—lots of bridges—complete the layout. As shown in the track plan, all but one of the bridges (excluding the mine spur bridge) are adjacent to O-31 sections of track.

To pull off this track plan, you have to choose your bridges carefully to keep locomotives and rolling stock from sideswiping the approaches. Ditto for the six tunnel portals, and the vertical walls in the aforementioned areas.

One spot on the layout—where the upper reverse-loop track, following a counterclockwise direction, crosses the river for a second time on its way to the central mountain peak—is absolutely perfect for a wooden trestle with a cutout for the mainline track to pass below.

In addition to the three coal-moving accessories mentioned, this track plan uses an American Flyer no. 789 baggage smasher station and a Lionel no. 464 operating sawmill. Lionel recently introduced reproductions of both accessories. Lionel nos. 138 water tower and 356 operating freight station round out the accessories.

Did I just say "round"? Build a layout based on this track plan, and your trains will most definitely be coming 'round the mountain.

You don't usually see outcroppings like this on a toy train layout, but one would work for this track plan. Burlington Northern diesels lead an eastbound coal train past Finger Rock, Colo.
John S. Murray

Scenery and accessories

Scenery will make or break a layout built with this track plan. Think vertical—and then add another foot or two in height. I struggled when creating the scenery on this track plan to get it steep enough!

The peak over the center of the three tunnels on the upper loop should be taller than the roofline of the Lionel no. 98 coal bunker, which for our purposes is the loading spot for the ore mine. That calls for near-vertical walls along the ore-mine spur.

Likewise, walls are rock-climber steep along the lowest section of the upper-reverse loop, where five pieces of O gauge track parallel the narrow stream. Think of it as a mini Grand Canyon carved by the

Prairiewood

What did you have for breakfast this morning? Oatmeal? Bagels? Frosted

Flakes? Chances are the grains used to make your breakfast came from middle

America. This track plan depicts just such a place: the wheat and cornfields of

the upper Midwest and the Great Plains.

This single-level O gauge track plan fits within a 9- by 12-foot space. There are no grades, but there are opportunities for landforms—note the elevated farm inside one of the loops of track.

On the far side of the layout is the edge of a small town. Elsewhere, the flatness of the prairie is broken up by a pair of Lionel grain elevators and, in this depiction, a few gentle berms to divide the scenes.

Visually, this track plan is a mix of extreme horizontal and vertical elements—amid amber waves of grain. And that's the prairie.

A train on this track plan wanders past the elevators, a town, and a farm. Two pairs of track switches adjacent to the central crossing allow the lobes of this irregular figure-eight to double as reverse loops. Trains can run in a clockwise direction, switch to counterclockwise operation, and return to a clockwise path.

Amber waves of grain frame a lonely Farmrail train near Lone Wolf, Okla. *Hal Miller*

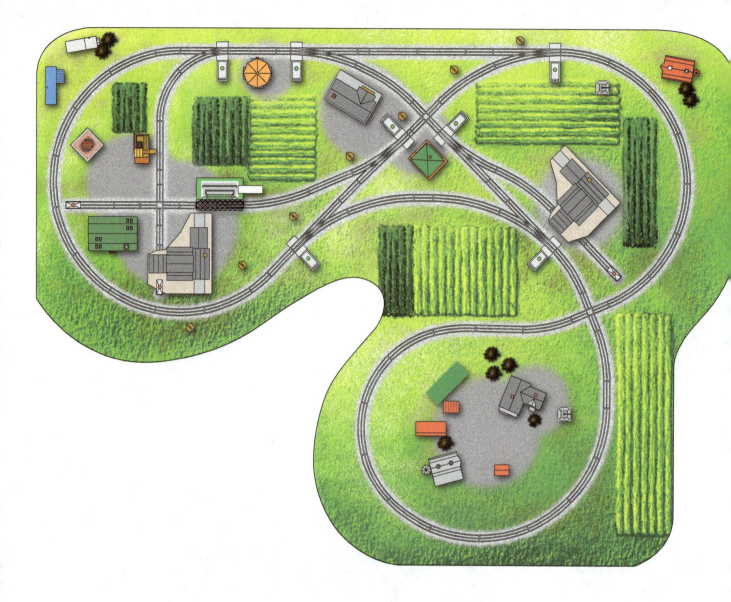

Track and wiring

As depicted, the track plan features regular Lionel O gauge track and O-42 and O-54 curves. It can be built with other brands of track offering O-54 curves.

Track switches are O-72, except for a single spur line that uses an O-31 switch. The three crossings are 90-degree sections. There are a handful of fitter sections on the plan near each of the crossings, but otherwise tracklaying is straightforward.

In the town, two spurs come from opposite directions and cross one another, adding visual and operating interest. If you wish, there's enough space for a short third spur inside this loop, the largest of the three on the layout.

At the farm end of the layout, the trackwork can be expanded, stretching the loop to an oval to allow for more spur lines and

another town. A passing siding along the straight section of the main line near the passenger station will be a welcome addition to the scene.

The wiring plan for conventional-control operation gives you the opportunity to hold a train at the loop around the farm while switching the grain elevators and small factory in town. The main line has two blocks; the central crossing has its own block; and each of the sidings is a block.

Depending on the brand of track you use, you may need to modify the crossings to keep the north-south and east-west lines electrically independent from each other. This is important where the two sidings in town cross each other. On Lionel O and O-27 crossings, cutting away part of the metal plate beneath the track section and adding jumper wires will keep each line's center rail electrically independent.

Above: A hood unit in a paint scheme reminiscent of the old Rock Island pulls a handful of freight cars through the brush near Mitchell, South Dakota. *Cody Grivno*

Madrid, Iowa, circa 1947. The Milwaukee Road tracks may as well have had three rails. In this scene are two grain elevators similar to Lionel's modern-era towers, a station not unlike MTH's Fairview depot, and, off in the distance, is that a postwar Lionel no. 193 industrial water tower? *Henry J. McCord*

Freight and passenger trains

At the heart of the figure-eight is a central crossing and the town's station. If you're an operator who favors a variety of road names, you can envision this spot as a crossing of perhaps the Rock Island and Chicago & North Western, or the Burlington Route and Soo Line, or the Santa Fe and Frisco. The combinations, real or imagined, are many.

Freight trains will be heavy with grain hoppers, but coal from the western ranges and container cars from Pacific ports can also be seen moving eastward. Livestock cars and boxcars with goods from small industries will be at home, too.

While you won't find a grand passenger terminal out here in the fields, some of the smartest looking passenger trains ever built crossed the plains. Gleaming Burlington

ACCESSORIES

QTY.	DESCRIPTION
1	Gilbert 772 water tower
1	Lionel 138 water tower
1	Lionel 3656 operating stockyard
1	Lionel 445 automatic switch tower
1	Lionel 12706 barrel loader building kit
2	Lionel 12726 grain elevator kit
1	MTH 30-9001 farmhouse (gray and yellow)
1	MTH 30-90005 stainless mobile home
1	MTH 30-90026 granary
1	MTH 30-90032 Fairview depot
7	Plasticville 1000 telephone pole
2	Plasticville 1408 windmill
1	Plasticville 1504 mobile home
1	Plasticville 1601 barn
1	Plasticville 1617 chicken coop and corn crib
1	Plasticville 1622 dairy barn with silo

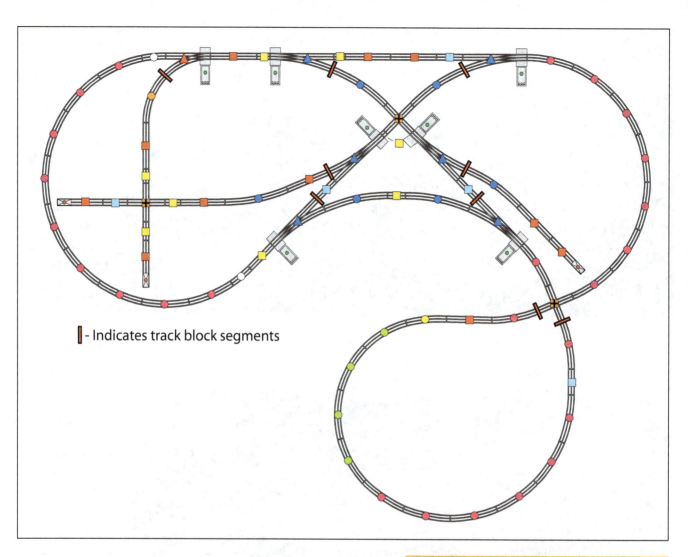

| - Indicates track block segments

Zephyrs and silver-and-red Rock Island Rockets were among the speed demons.

On the other side of the passenger timetable were locals, serviced by old Alcos pulling a mail car and a single coach, or doodlebugs, like those made by MTH (doodlebugs look like motorized heavyweight passenger cars).

Scenery and accessories

Key to this track plan is the depiction of crops. There are a few ways to avoid having to plant hundreds of stalks of 1:48 scale corn. German model railroad manufacturers offer sheets of scenic paper with textures that mimic rows of green or golden crops in various stages of growth or even a freshly plowed field. (These commercially available sheets are far more sophisticated than the sawdust grass mats we used as youngsters.)

Trimmed and bordered with fences or wild grasses, the appearance of this scenic paper is realistic. If your hobby shop doesn't carry these products, try suppliers such as Wm. K. Walthers (www.walthers.com) or Scenic Express (www.scenicexpress.com).

LIONEL TRACK

QTY.	DESCRIPTION	
11	Lionel single straight	🟧
5	Lionel half-straight	🟦
8	Lionel cut straight	🟨
1	Lionel 0-31 curve (45 degrees)	🟠
4	Lionel 0-42 curve (30 degrees)	🟢
25	Lionel 0-54 curve	🔴
2	Lionel 0-54 half-curve	⚪
1	Lionel 0-54 cut curve	🟡
6	Lionel 0-72 curve (22.5 degrees)	🔵
1	Lionel 0-31 modern left-hand turnout	🔺
3	Lionel 0-72 left-hand turnout	🔺
3	Lionel 0-72 right-hand turnout	🔺
3	Lionel 90-degree crossing	✚
3	Lionel 260 track bumper	

Rock Island GP38-2s, in their final light blue paint scheme, stand out amid the earth tones of the prairie. *David P. Morgan*

Other crop options include indoor-outdoor carpeting and plastic artificial turf, the green or brown type you often see used for doormats. The key to using these materials is color. Try a light overspray from a paint can—yellow and brown if you're trying to make a green mat look like a young cornfield. Follow that with a coat of non-glossy clear (crops aren't shiny!).

Even though you may never fully disguise the origins of your field, experimenting with oversprayed colors will keep visitors from commenting about "Astroturf farms."

In town, streets and structures should look dusty and sparse. A clever addition will be a farm implement dealer, giving you a wonderful excuse to round up a dozen or so die-cast metal pieces of farm equipment.

Horizontal structures and accessories can temper the dominating grain elevators. On this track plan there's a postwar Lionel operating stockyard, along with some nondescript, low-lying structures.

On the farm end of the layout, the barn and outbuildings are Plasticville and the farmhouse is MTH. All are closer to S scale than O. But when raised a few inches above track height, they help you create the illusion that there's enough room for a farm within a loop of O-54 and O-42 track sections.

And that's the key to this track plan—creating the illusion of distance to model the great prairie in just a 9- by 12-foot space.

EIGHT

Breaking the Rule of One

Most toy train layouts feature one water tower. One log loader. One automatic gateman. One tunnel. One factory. One banjo signal. They are bound by the "rule of one." While few of us have just one locomotive, the rule of one, nonetheless, still applies. One postwar Lionel Berkshire. One GG1 electric locomotive. One Lionel Phantom locomotive. One operating milk car. One bay window caboose.

H ere's a walk-in track plan for a 12- by 12-foot space that breaks free from that pesky rule of one. There are two ovals for two-train operation. There are two industrial wings: one a seaport and the other a mining operation.

At the port are two Lionel gantry cranes. At one end of the mining spur are two American Flyer Seaboard coalers, and at the other end there are—not just two—but *three* modern-era Lionel ore loaders.

I can't take credit for the overall shape and design of this track plan. A fellow

A red, white, and blue Central Vermont diesel leads a hopper train full of stones through late-summer foliage near Westminister, Vt., in 1988. *John S. Murray*

named Kenneth Gentili designed a track plan for the fictional Spokane, Pasco, & Wallace Railroad, and it was published in the September 1992 issue of *Model Railroader* magazine. His HO scale track plan, more complex than the one shown here, was designed to make the most of a corner in a room. Likewise, this O gauge track plan is designed to fit into a corner.

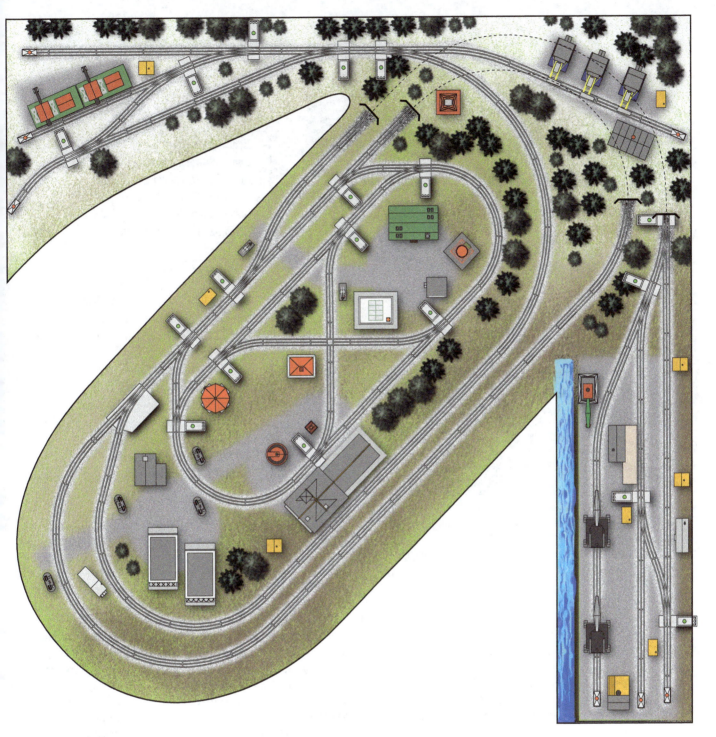

Track and wiring

As shown, this track plan uses Lionel tubular track and Lionel and K-Line track switches, but it can be built with any brand of track offering O-31, O-42, O-54, and O-72 curves. To build it as shown, you'll also need O-31, O-42, and O-72 track switches.

The main oval uses O-54 curves, as does the approach track to the port. Both lines burrow through two tunnel portals at the rear corner of the layout.

A secondary oval that doubles as a figure-eight is within the main oval. It uses O-31 curves and track switches, and it allows trains on the main oval to reverse direction to run in a clockwise or counterclockwise direction.

The mining spur departs the main oval through an O-42 track switch (it's the only O-42 switch on the layout; a switch of a different size can be adopted) and swings around the main oval more than 270 degrees while climbing up and over two of the tunnel portals.

How's this for low-level, toy train tunnels? The State Line Tunnel in New York looks just right for the main line and one of the spurs on this track plan. John Nehrich

Shipping ports and trains are no strangers. An Alco road switcher belonging to the Atlanta & St. Andrews Bay Railway pulls a pair of tank cars along the rails in Panama City, Fla., in 1953. W. V. Anderson

The spur includes a crossover track to allow a switcher to service the two coal loaders and the three ore loaders at the far corners of the layout. There's also a crossover track along the port spur.

To wire this track plan for conventional-control operation, you'll need to break the main oval into three blocks to allow you to move a train in and out of the inner O-31 loop while a second train waits. The long approaches to the mining spur and the port spur should be separate blocks, as should each of the tracks at the coal and ore loaders and the gantry cranes.

For two-train operation, transformers with two throttles, such as MTH's Z-4000, MRC's Dual Pure Power, and Lionel's ZW (postwar or modern), are ideal. With so much action on the rails, I recommend using an auxiliary transformer for the lights and accessories.

Freight and passenger trains

Depending on your choice of accessories, this track plan has the flexibility to be many railroads. If your mining spur

accessories are all coal, look toward railroads like the Virginian, Norfolk & Western, and Chesapeake & Ohio, which hauled coal to ports on the Eastern Seaboard.

Not hot about Eastern roads? The Great Northern hauled ore to Lake Superior and at one time had the largest ore docks in the world. Imagine a pair of six-axle diesels or a weather-worn steamer barreling down the main oval and turning into the port spur, with two dozen of the short ore cars made by Lionel, K-Line, and MTH in tow.

Don't like hauling ore? Change the cargo. The mining spur can accommodate factories or warehouses, and the waterfront can be a shipping port on the Gulf of Mexico serviced by boxcars from the Southern Pacific, Missouri Pacific, Illinois Central, or Kansas City Southern. You can also turn this track plan into a lumber hauler, taking harvests down to the waterfront for shipment across the seas.

While the focus of this track plan is clearly freight, a diesel streamliner or a steam locomotive, tugging away at a string of heavyweight coaches, can be rounding the main oval while freight switchers work both

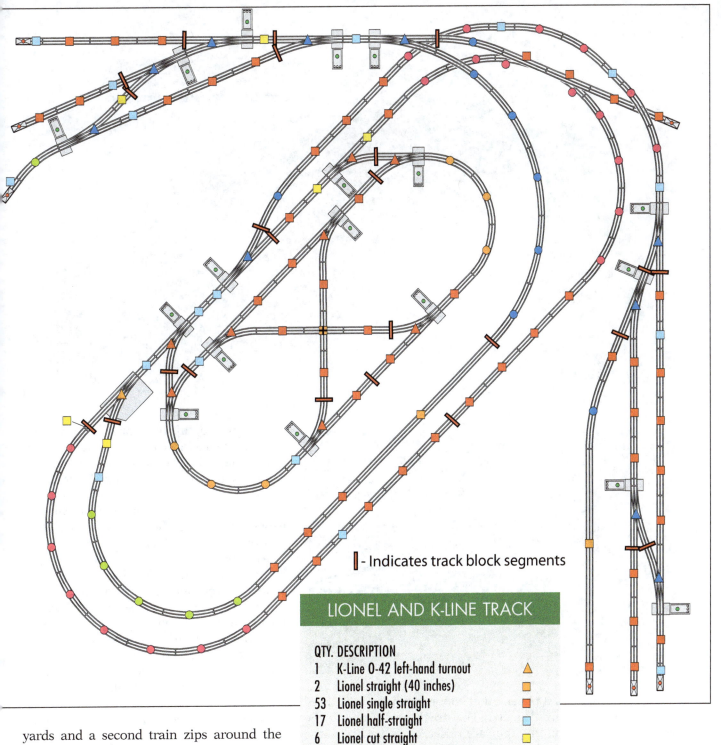

| - Indicates track block segments

LIONEL AND K-LINE TRACK

QTY.	DESCRIPTION	
1	K-Line O-42 left-hand turnout	🔺
2	Lionel straight (40 inches)	🟧
53	Lionel single straight	🟧
17	Lionel half-straight	🟦
6	Lionel cut straight	🟨
6	Lionel O-31 curve (45 degrees)	🟠
6	Lionel O-42 curve (30 degrees)	🟢
22	Lionel O-54 curve	🔴
8	Lionel O-72 curve (22.5 degrees)	🔵
4	Lionel O-31 modern left-hand turnout	🔺
4	Lionel O-31 modern right-hand turnout	🔺
8	Lionel O-72 left-hand turnout	🔺
2	Lionel O-72 right-hand turnout	🔺
1	Lionel 90-degree crossing	✚
7	Lionel 260 track bumper	

yards and a second train zips around the O-31 figure-eight.

Scenery and accessories

Twin tunnels are the biggest scenery feature of this track plan. One allows a train to hide from view as it rounds the back end of the main oval. The second tunnel quietly connects the port spur to the main line. Because this track is hidden, it makes the port seem much farther away.

Above these two tunnels is a relatively

Can you count how many ore cars are in this photo of the Duluth, Missabe & Iron Range Railway? I can't, but this track plan can handle plenty of these short freight haulers.
Jim Hediger

As depicted, this track plan features doubles and even a few triples when it comes to accessories. The waterfront has two Lionel gantry cranes and an American Flyer fixed-based crane. A double dose of Plasticville hobo shacks and a railroad workcar are peppered between the tracks, along with a pair of Lionel shed kits.

At the left end of the ore spur are a pair of towering American Flyer Seaboard coalers; at the right is a trio of modern-era Lionel ore loaders. With so many different loaders available today—both vintage and modern—substitutions are welcome.

The central part of this track plan is a bit on the eclectic side. Shown are two MTH RailKing textile factory buildings (the fire escapes on the backsides are first rate), a modern-era Lionel Rico passenger and freight station, an MTH RailKing industrial building, and MTH tinplate reproductions of the prewar nos. 436 power station and 438 switch tower.

Two pairs of Lionel crossing signals and two banjo signals protect the two roads passing through town. The list of doubles just never seems to end.

Finally, in the woods beyond the Rico station, there are three French hens, two turtledoves, and a partridge . . . oops, just kidding. But as you can see from this track plan, the "rule of one" is meant to be broken.

flat plateau that contains a line to the three ore loaders. The tracks throughout the mining spur are raised one level above the main portion of the layout and the waterfront spur.

Scenery here can be a bit on the rugged side to suggest that these raised areas of the layout are in a different geographical region from the lower levels.

At the port, water is kept to a minimum—just a strip along the dock is enough to suggest a large body of water beyond the edge of the layout. Make it a bit wider, though, and you can include one of Lionel's new animated tugboats.

The Quiet One

It's not the loud ones, but the quiet ones that you need to pay attention to.

This track plan is a quiet one—there are no spectacular trestles, triple-track main

lines, or tiered crossovers. But look a little deeper and you'll find gentle curves,

an active 90-degree crossing, and two industrial areas.

Look even closer and you'll see a single-level plan with a double-duty reverse loop for clockwise and counterclockwise operation, four short bridges over two creeks, and an industrial yard with a forked approach track.

Track and wiring

This O gauge walk-in track plan fits into a 9- by 12-foot space. It features regular Lionel O gauge track, including ten O-72 track switches. There are also a pair of K-Line O-42 track switches. Both return loops use O-42 curves. A smattering of O-54 and O-72 curve sections tie the layout together.

You can easily substitute K-Line Shadow Rail track for the sections shown on the drawing. Or find another brand that offers the same diameter curve sections.

Sharpen your saw blade before you get started, however, since you'll have to cut 11 straight sections to lengths ranging from 1½ to 7¾ inches. Most of these sections are used to accommodate the 90-degree crossing and the paired crossover switches along the back side of the layout.

Cutting tubular track is easy. Mark your cuts, sandwich your track between two blocks of wood to keep the rails from shifting, put the whole thing in a bench vise, and slowly draw your blade back and forth.

This task is even easier if you use a motor tool, like those made by Dremel or

A Conrail GP40-2 diesel is on the point of a freight train passing through a quiet yet scenic area of upstate New York. *Jim Wakeman*

Black & Decker. Go slowly, however, and use a reinforced cutoff blade. Absolutely don't forget to put on your safety glasses.

Conventional-control wiring for this track plan focuses on the two industrial areas. For the peninsula yard, each track should be its own block.

To facilitate locomotive movement, split the approach track into two blocks: first, the direct route along the right edge of the layout; second, the alternate route that includes the 90-degree crossing. You also need to electrically isolate the center rail of the north-south and east-west routes of the Lionel crossing section by cutting its metal contact plate and adding jumper wires.

On the other end of the layout, make the siding its own block. Do the same thing to the reverse loop section beyond the crossover track switches.

Freight and passenger trains

This is a quiet track plan, and that's good. No Big Boys, no Veranda Turbines, and no 45-car trains. This track plan lends itself perfectly to postwar Lionel trains and their modern-era kin from MTH, K-Line, Williams, Weaver, and of course Lionel.

The plan reminds me of a secondary line in eastern Pennsylvania, upstate New York, or parts of Ohio. Small mixed-freight trains led by postwar Lionel 2-6-2 or 2-6-4 steamers or New York Central and Pennsylvania Railroad F3 diesels will be at home here.

And don't forget plenty of new and old gondolas. You'll need them with the Lionel culvert loader, culvert unloader, and barrel loader on the track plan.

Doodlebugs, Rail Diesel Cars, or short strings of heavyweight coaches will be sufficient for passenger service. Or you can have a streamlined name passenger train "just passing through."

Scenery and accessories

Like trees? There are plenty to plant on this layout. When you think you've planted enough, add more. Inside both mainline

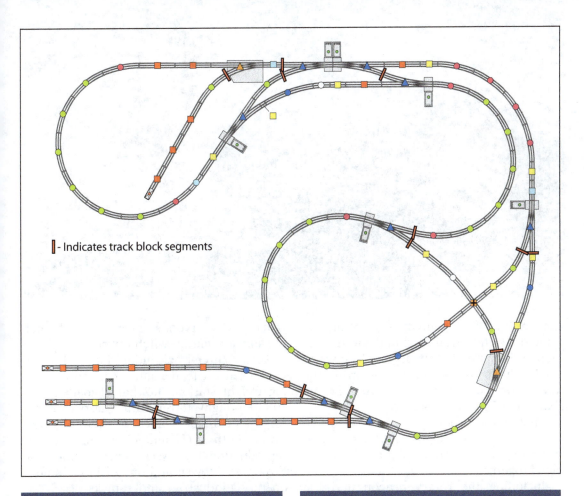

- Indicates track block segments

LIONEL AND K-LINE TRACK

QTY.	DESCRIPTION	
2	K-Line 0-42 left-hand turnout	▲
26	Lionel single straight	■
2	Lionel half-straight	■
11	Lionel cut straight	■
24	Lionel 0-42 curve (30 degrees)	●
9	Lionel 0-54 curve	●
1	Lionel 0-54 half-curve	●
4	Lionel 0-72 curve (22.5 degrees)	●
3	Lionel 0-72 cut curve	○
3	Lionel 0-72 left-hand turnout	△
7	Lionel 0-72 right-hand turnout	△
1	Lionel 90-degree crossing	✚
4	Lionel 260 track bumper	

ACCESSORIES

QTY.	DESCRIPTION
1	Gilbert 755 talking station
1	Lionel 138 water tower
1	Lionel 199 microwave relay tower
5	Lionel 314 plate girder bridge
1	Lionel 342 culvert loader
1	Lionel 345 culvert unloader
1	Lionel 356 operating freight station
1	Lionel 362 barrel loader
1	Lionel 395 floodlight tower
1	Lionel 445 automatic switch tower
1	Lionel 494 rotary beacon
1	Lionel 12718 barrel shed kit
1	Lionel 12733 watchman's shanty kit
1	Lionel 12944 Sunoco oil derrick
1	Lionel 14142 operating industrial smokestack
1	MTH 30-9126 Flyer 23780 Gabe the Lamplighter
1	MTH 30-11051 Flyer 23772 water tower
1	MTH 30-90001 outhouse
1	MTH 30-90026 granary
2	Plasticville 1627 hobo shack

loops are gentle hills punctuated by a Lionel no. 494 rotary beacon on one mound and a Lionel no. 199 microwave relay tower on the other. The tree-covered hills partly obscure trains as they pass behind, creating a greater sense of distance.

Beyond green foliage are a pair of creeks at both ends of the layout. They visually

A Chessie System GP9 diesel passes Viaduct Junction, Md., in August 1984. Give the tower a green roof (and two blue men!) and you'll be looking at a postwar Lionel no. 445 automatic switch tower. *Don Hodon*

expand the track plan by separating the three-track industrial yard and the reverse loop at the far end of the layout. The water can be blue paint, or you can use one of the new water-creation products from Scenic Express or Woodland Scenics.

That's it for scenery. Since all of the track in this plan is on a single level, there are no tricky tunnels or sections of graded track to blend together.

Including the rotary beacon and the microwave tower, there are more than a dozen operating accessories on this track plan. The single spur at one end of the layout features a Lionel culvert unloader and oil derrick. In the center are a Lionel no. 138 water tower, an American Flyer no. 755 talking station (Lionel reissued it just last

year), and everyone's favorite, the Lionel no. 445 automatic switch tower.

In the three-track industrial yard are a Lionel culvert loader and nos. 356 operating freight station, 362 barrel loader, and 395 floodlight tower. You'll also find an American Flyer Gabe the Lamplighter tower (both MTH and Lionel have made reproductions), an MTH reproduction of a Flyer water tower, and one of Lionel's new operating industrial smokestacks.

Rounding out the picture are an MTH RailKing industrial building, outhouse, and Plasticville hobo shacks. Phew! All of those great structures on an unassuming 9- by 12-foot layout!

As I said at the outset, it's the quiet ones that you need to watch.

Through the Woods

Less can be more, even when it comes to toy train layouts. Here's a walkaround track plan designed for two-train operation. It fills a 12- by 12-foot area, but has only eight structures. Eight? In a 12-foot-square space?

It's a foggy fall morning as this eastbound grain train dashes through the woods near Foley, Pa., along the famed Sand Patch line. *Mark Perri*

Yes! Look closely. Real railroads go places; they don't simply circle endlessly through town. This track plan features an oval using O-54 curves and a passing siding accompanied by a wye connecting the main body of the layout to a flowing up-and-over reverse loop.

As depicted, the reverse loop wanders through the spaces between towns and cities. When I was a kid, we called those spaces "The Woods." There's also a river to cross and several modest bridges—nothing more. That's the beauty of this less-is-more plan.

Track and wiring

At the heart of the track plan, which relies on tubular straight track and O-42, O-54, and O-72 curves, is a basic oval. A train, running on the oval in a counterclockwise direction, passes through one leg of the central wye as it ducks beneath the reverse loop, heading for the passing siding at one end of the track plan.

After rounding the 180-degree bend, our train passes through a small town with lumber-themed accessories before entering another 180-degree curve. The entire main line uses curves no tighter than O-54, so it can accommodate all but the largest premium-priced O gauge steam locomotives.

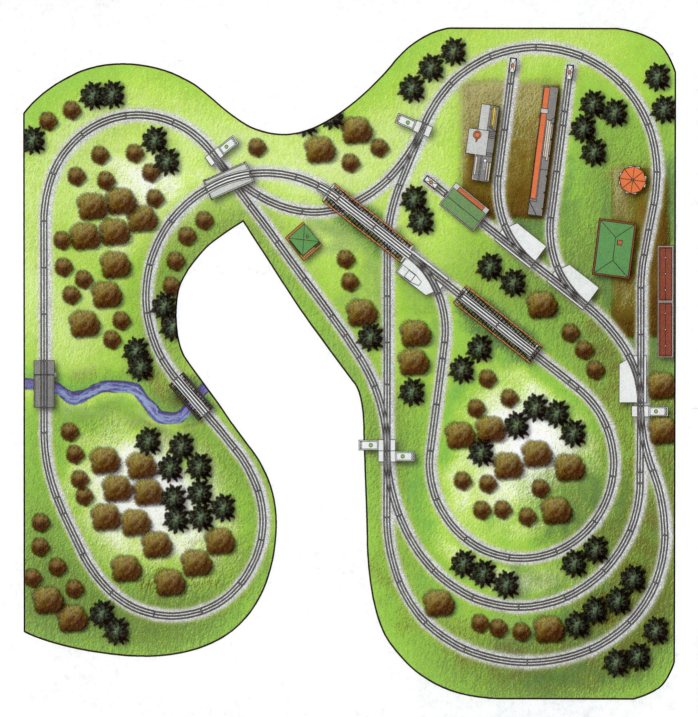

Taking the diverging route at the wye, our train turns onto the long, scenic reverse loop. Here the train reverses direction four times as it climbs up and over the wye before reaching an O-42 track switch. Then it returns down the same winding grade to the main oval, where it can continue in its original direction of travel or reverse itself.

Even in conventional-control mode, you can keep two trains running on this track plan. Just remember that, as dispatcher, you have to play an active and not a passive role.

While one train follows the oval, a second train can make its way through the winding

reverse loop. If you break the wye into separate electrical blocks, the train on the oval can enter the wye, stop, and wait while the second train clears the reverse loop, takes an alternate path through the wye, and enters the main oval. The first train can then move forward into the reverse loop so you'll again have two trains under way.

This realistic switcheroo can be repeated as often as you like. Also, by utilizing the passing siding on the main oval, a third train can weave in and out of the action.

In addition to the blocks for the wye approaches and the passing siding, you'll

ACCESSORIES

QTY.	DESCRIPTION
4	Gilbert 571 truss bridge
1	Lionel 314 plate girder bridge
1	Lionel 497 coaling station
1	Lionel 2324 operating switch tower
1	Lionel 12812 illuminated freight station
1	Lionel 12916 "138" water tower
2	Lionel 12943 illuminated station platform
1	Lionel 14001 "364" conveyor lumber loader
1	Lionel 14173 drawbridge
1	Lionel 32989 "464" operating sawmill
1	Lionel 62716 short extension bridge

want to create separate blocks for the operating accessory tracks.

Freight and passenger trains

With a pair of lumber-themed accessories, you'll need plenty of wood-hauling flatcars on this track plan. I've also carved out a spot for a postwar Lionel no. 497 coaling station (Lionel produced a smooth-running modern version of this accessory just a few years ago), so operating dump cars also are the right ticket for this track plan.

Beyond that, mixed freights are in order, whether comprised of postwar Lionel cars or modern-era cars from MTH, Lionel, K-Line, Atlas O, and Williams. Whatever you favor will work on this track plan.

The same goes for motive power. The plan shows an operating water tower, and, if you favor Lionel or MTH's walkaround controllers, there are ample spots on all four sides of this track plan for steam railfanning. It's exciting to envision one of the new MTH RailKing steamers in labored chuff and smoke-output modes snorting its way up and around the wooded reverse loop, twice crossing a river and the wye tracks below.

Passenger service can consist of through streamliners or local service with MTH doodlebugs; Lionel, MTH, or K-Line Rail Diesel Cars; or timeworn steamers pulling a pair of heavyweight coaches and a combine over the river and through the woods.

If steam isn't your game, ditch the water tower and lash up a pair of Geeps or an A-B-A cab-unit combo (Alcos or EMDs), or mate contemporary EMD or GE diesels to your freight train. (Right now I'm envision-

ing Lionel's new LionMaster Conrail SD80MAC diesel with blue and white flanks pulling a brace of freight cars through the golden autumn foliage—heck, even my non-railroady wife would find that scene appealing!)

Scenery and accessories

This is a track plan built for big patches of trees. I've selected autumn trees with a few firs thrown in, but you can just as easily model spring or summer (but don't dismiss autumn—in your train room it looks as appealing in February and July as it does in October).

You can do this plan's trees two ways—expensive or cheap.

If you have the money, by all means spend it on realistic trees. Outfits like Scenic Express and Woodland Scenics offer plenty of choices. If you don't want to buy 250 or more trees (that's what it will take to do this layout justice), turn to nature.

Here's the game plan. Use trees from Scenic Express, Woodland Scenics, and other suppliers in the foreground. Use clumps of dyed ground foam, available

Top: A Conrail General Electric B23-7 diesel crosses a small bridge amid fall foliage near Alfred Station, N.Y., in October 1986. Note the Plasticville-like "hobo shack" at the base of the bridge. *Charles Woolever*

A Conrail train passes along the edge of what can only be a Lionel log-loading and unloading accessory, of which there are two on this track plan. *Scott Hartley*

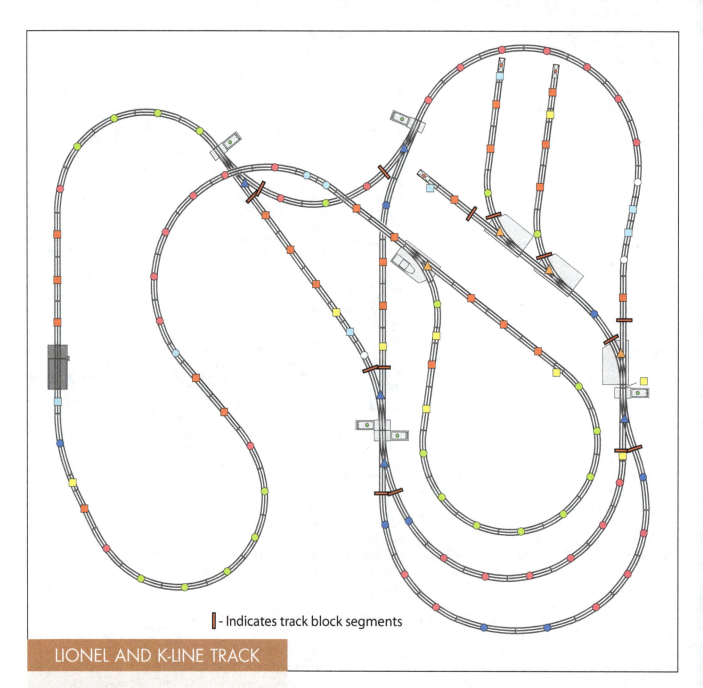

▮ - Indicates track block segments

LIONEL AND K-LINE TRACK

QTY.	DESCRIPTION	
1	K-Line O-42 left-hand turnout	▲
3	K-Line O-42 right-hand turnout	▲
25	Lionel single straight	■
6	Lionel half-straight	■
8	Lionel cut straight	■
21	Lionel O-42 curve (30 degrees)	●
29	Lionel O-54 curve	●
3	Lionel O-54 half-curve	●
8	Lionel O-72 curve (22.5 degrees)	●
3	Lionel O-72 cut curve	○
4	Lionel O-72 left-hand turnout	△
1	Lionel O-72 right-hand turnout	△
3	Lionel 260 track bumper	

from the same suppliers, for trees farther back in the woods. Behind the front layer of trees, you don't see the trunks and branches anyway, just the treetops. So why build or buy all of those trunks?

Another way to fill the forests is with help from Mother Nature. Trees can be modeled from a variety of dried weeds or twigs, accentuated with pieces of ground foam in spring green or autumn gold colors.

No weeds in your yard? No problem. Dried plants in a wide variety of shapes and colors—many scaled for O gauge—are sold in craft shops.

Don't be afraid to walk into a craft shop— just play dumb and pretend you're on a

From a distance, it looks like a truck-load of Woodland Scenics trees was used to create this scene along the Hudson River at Bear Mountain, N.Y. *Ken Karlewicz*

mission for your spouse if any sales clerks ask if you need help. And be sure to visit these stores during different times of the year—you're more likely to find spring foliage colors in March and autumn colors in September.

For more information on modeling trees, thumb through some of the model railroad scenery books available from Kalmbach Publishing Co. Just because some books are focused on HO scale railroads doesn't mean the scenery techniques they describe won't apply to toy trains.

Beyond the forest, there are some carefully selected toy train accessories on this track plan. In addition to the previously mentioned Lionel operating sawmill, log ramp, and coaling station, I've chosen an operating water tower, an illuminated passenger and freight station, and, on the side of the layout farthest from the town, one of Lionel's new small operating drawbridges.

There also are four American Flyer no. 571 truss bridges on the track plan (wide enough for O gauge trains, their open design is a welcome alternative to solid girder bridges). You can substitute other bridges to meet your tastes.

The theme of this track plan is "less is more." So how did this chapter, about a layout with only eight structures, wind up being among the longest in this book? Because sometimes, when it comes to toy train track plans, less really is more!

The Works

Sometimes you just have to be bold and even a little brash. A jumbo pizza with everything—including extra anchovies. A 44-inch wide-screen TV—no, make that a 52-inch—with a home-theater sound system. Or how about a 12- by 12-foot O gauge layout with four of Lionel's largest tinplate accessories?

This track plan is no wallflower. Lionel's biggest passenger station sits atop a landscaped terrace, and two complete loops (one shaped like a boomerang) are perpendicular to each other. Two pairs of crossover switches, six crossing sections, and two industrial spurs complete this 12- by 12-foot layout.

Track and wiring

Designed with Lionel O gauge track, this plan has two loops at its core. One is a traditional oval with O-42 curved sections on

One example of "The Works" is the Southern Pacific's no. 4449 locomotive in its brilliant Daylight paint scheme. Lionel, MTH, K-Line, and Williams have made die-cast metal O gauge models of this locomotive. *Jack D. Kuiphoff*

each end. The other loop, running perpendicular to the first, uses O-42 and O-54 curves and features a long passing siding on one end.

Inside the oval loop is a forked industrial spur with three operating accessories. Inside the boomerang-shaped loop is another spur with three more accessories. This spur has

an extended approach track that passes across the front of the Lionel City Station, intersecting the first loop twice.

The two loops cross each other in four places—six if you count the industrial spur. Trains can pass between the loops using two pairs of O-72 switches that act as crossovers. The crossovers create a long run if a train swings back and forth between loops as it moves through the layout. All track switches are O-72, except for a lone O-31 switch within the oval loop.

For conventional-control operation, the track plan presents two wiring choices.

The first (and easier) path calls for separate blocks for the accessory sidings, the long approach spur, and the two loops. You'll want to make the passing siding on the boomerang-shaped loop and its adjacent main line separate blocks to facilitate moving trains in and out of two loops.

The second level of wiring involves automatic train protection at the four mainline crossings when operating two trains at once, each on its own loop. Automatic train protection means the approach sections to all the crossings will be individual blocks. When a train passes

QTY.	DESCRIPTION
1	Gilbert 583 electromagnetic crane
1	Gilbert 586 wayside station
1	Lionel 45 automatic gateman
1	Lionel 092 signal tower
1	Lionel 116 station
1	Lionel 126 station
1	Lionel 127 station
1	Lionel 129 terrace
1	Lionel 137 passenger station
10	Lionel 153 automatic block signal
1	Lionel 397 diesel-type coal loader
1	Lionel 436 power station
1	Lionel 438 signal tower
1	Lionel 840 industrial power station
1	Lionel 12915 "164" log loader
6	Lionel 12943 illuminated station platform
1	Lionel 14134 "182" magnetic crane
1	Lionel 22918 locomotive backshop
1	Lionel 32905 Lionel factory
1	Lionel 49807 Flyer 752 Seaboard coaler

Top: Prewar Lionel tinplate accessories weren't purely the fantasy of Lionel's product designers. You can see Lionel's nos. 115 and 116 city stations in the lines of this Southern Pacific station in San Jose, Calif. Southern Pacific Co.

Bottom: Lionel's no. 438 switch tower clearly shares the pagoda-style roof and framework of this tower in Winchester, Ind. Linn Westcott

through the crossing, it turns off power to the intersecting route's block, preventing a possible collision. The track plan shows track-signal placements for such an arrangement.

Should you choose this second level of wiring, I suggest you lengthen the size of the oval loop. Otherwise, trains in that loop will be stopping more often than they're going.

Freight and passenger trains

Speaking of trains, this track plan calls for your best. Keep your ratted-out gondolas in the scrap yard near the Lionel modern-era locomotive backshop accessory.

The top-drawer tinplate accessories used here call for some snazzy streamliners like the prewar Lionel Flying Yankee and American Flyer O gauge Union Pacific City of Denver; postwar trains like an F3-led Santa Fe streamliner set or GG1 Congressional set; or modern-era speedsters like K-Line's Reading Crusader and MTH's RailKing Hiawatha and Southern Pacific Daylight sets.

Don't forget streamlined diesels, like Alco PA-1s from Lionel and MTH, as well as Electro-Motive E units from Williams, Weaver, and K-Line. There's also the brand-new Hiawatha service Fairbanks-Morse Erie-Built from Atlas O.

Freight service has a role, too. An old warhorse steamer can service the scrap yard and its gondola loads, while on the oval loop there's a need for flatcars to carry lumber and hoppers to carry coal. And someone has to fuel that giant no. 840 power station!

Scenery and accessories

While the accessories shown on the track plan are decidedly tinplate, not all of them have to be old. The grand Lionel no. 116 city station has been reproduced in modern times by Lionel and others. Same for the no. 129 terrace that's used as a base for the station. Across the front of the station are two rows of Lionel station platforms, which have been in production on and off for more than half a century.

At the north end of the boomerang loop is a prewar no. 840 power station, among the largest accessories Lionel ever made.

Any O gauge track plan entitled "The Works" needs to include a model of the Jersey Central's Blue Comet. This two-tone blue beauty, an all-time toy train favorite, sped between Jersey City and Atlantic City in the years before World War II. *Don Wood*

MTH and Lionel have made modern-day reproductions of this behemoth.

Reissues are available for a number of other prewar and postwar Lionel accessories dotting the layout: the nos. 164 log loader, 282 gantry crane, 397 diesel-type coal loader, 436 power station, and 438 signal tower. The American Flyer nos. 583 magnetic crane and 752 Seaboard coaler have also been reproduced. Even the Lionel no. 153 block signals can be modern-day products.

At the south end of the boomerang loop are two modern tinplate structures: the multi-story Lionel factory and the Lionel operating locomotive backshop. Not one but two magnetic cranes—one

More plausibility for this prewar-style track plan: a Long Island Railroad train passes a switch tower next to an electric company facility that is next to a track switch. The same positioning of tracks and structures can be found on this track plan. And how about that Lionel-style utility pole to the left? *Don Wood*

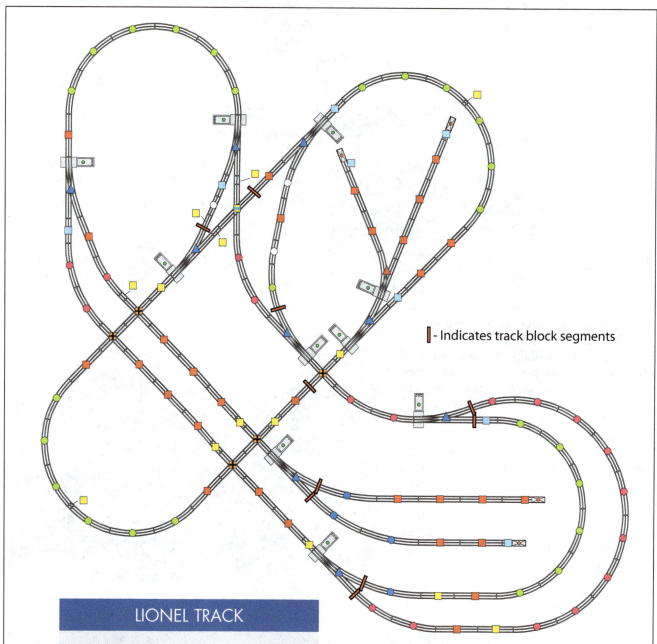

|| - Indicates track block segments

LIONEL TRACK

QTY.	DESCRIPTION	
30	Lionel single straight	🟧
8	Lionel half-straight	🟦
14	Lionel cut straight	🟨
25	Lionel O-42 curve (30 degrees)	🟢
18	Lionel O-54 curve	🟠
4	Lionel O-72 curve (22.5 degrees)	🔵
3	Lionel O-72 cut curve	⚪
1	Lionel O-31 modern left-hand turnout	🔺
7	Lionel O-72 left-hand turnout	△
2	Lionel O-72 right-hand turnout	△
1	Lionel 45-degree crossing	⊠
5	Lionel 90-degree crossing	✛
4	Lionel 260 track bumper	

Lionel and one Gilbert—provide operating enjoyment.

Even with all of these accessories, there's room for scenery. To accent the colorful accessories, I've shown a selection of autumn trees and a few fir trees.

There are enough trees to create the sense of a real landscape, yet not so many that they compete with the trains and accessories for attention. Not shown are roadways, which you'll need to display your collection of 1:43 die-cast metal vehicles (don't we all have too many of them?).

Any way you cut it, this track plan is the brashest in this book. It's got the works, even if you hold the anchovies!

Crossing the Ravine

As we followed the twists and turns of the multi-lane highway in our rented Chevrolet, headed toward Pittsburgh for a Train Collectors Association convention—wham, there it was. Looking too tall and too spindly, it was a railroad bridge—a mix of brown and black and rust cut diagonally above the highway between two wooded peaks.

A Norfolk & Western class A steamer roars across a multi-style bridge in Coopers, W. Va., during a historical society excursion run held in 1988.
Robert W. Lyndall

The words "Norfolk & Western" were clearly readable, and the first thing that came to mind was whether our timing would be right to see a N&W J-class locomotive race across the gap. Never mind that the locomotive and our rental car are nearly 50 years apart on a timeline. Seconds later, that bridge was behind us: out of sight, but not out of mind!

Track and wiring

While this track plan doesn't depict Pittsburgh with any accuracy, and the real N&W bridge isn't nearly as sexy as the MTH 30-inch truss bridge, this walkaround track plan is just as inspiring as the view was from our rental car.

The L-shaped layout, 5 feet across in most places, fits comfortably in a 9- by 12-foot space. The track used to create this plan is one of my favorite combinations— O-27 straights and wide-radius curves with Ross Custom Switches.

Say what? Up-market track switches combined with the cheapest straights and curves made? Yes. It's an economical way to have the best of both worlds.

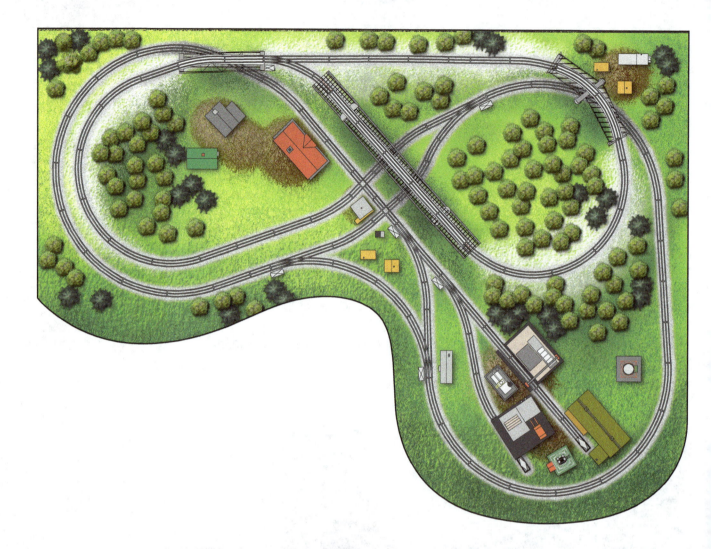

ACCESSORIES

QTY.	DESCRIPTION
1	Lionel 193 industrial water tower
1	Lionel 264 automatic forklift platform
1	Lionel 342 culvert loader
1	Lionel 12734 passenger/freight station
1	Lionel 12905 factory kit
1	Lionel 14143 industrial tank
1	MTH 30-9044 row house 1 (brown)
1	MTH 30-9126 Flyer 23780 Gabe the Lamplighter
1	MTH 30-90002 telephone shanty
1	MTH 30-90005 stainless mobile home
1	MTH 30-90010 workhouse 2 (gray and black)
1	MTH 40-1013 steel arch bridge (30 inches)
2	MTH 40-1014 girder bridge (10 inches)
1	Plasticville 1402 switch tower
1	Plasticville 1625 railroad workcar
4	Plasticville 1627 hobo shack

The money you save using O-27 track can be put toward Ross track switches. The rails of O-27 track and Ross track (also GarGraves and Curtis track) share the same rail height, so no shims are necessary. Sections of O-27 track, with a lower profile, can look more realistic than regular O track, especially when additional ties and ballast are used.

Adapter pins are readily available for O-27 and Ross/GarGraves/Curtis track. If you haven't already figured this out, O-27 track is produced in curve diameters of O-42 and O-54. There are no "O-27 curves" on this track plan—just O-42 and O-54.

The main level of this track plan is a figure-eight built with O-54 curves and track switches. One of the two reverse loops—trains can run clockwise and counterclockwise without backing up—is created by the gentle O-54 curve arcing between the two track lobes at the front of the layout.

The other reverse loop creates the most attractive element of this plan. The loop

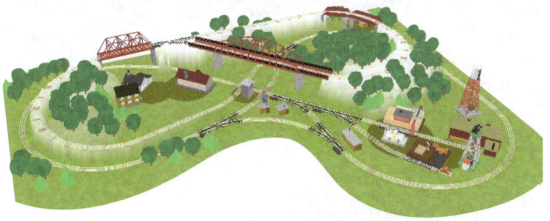

actually starts on the backside of the layout, running beneath the central bridge on one of the paired 90-crossing track sections.

The reverse loop then makes a sweeping right-hand O-42 turn as it climbs and crosses the main line below. At this point, an O-42 track switch takes trains either clockwise or counterclockwise across a short bridge at the far corner of the layout and the grand bridge—more than 4 feet of MTH truss bridge and girder bridges—before descending to the main line.

Wiring this track plan in conventional-control mode calls for independent blocks on the upper return loop and one, if not two, blocks for the industrial area, depending on whether you wish to store a switching locomotive there.

The main figure-eight should be broken into three blocks. That arrangement facilitates moving trains between the upper reverse loop and the figure-eight if you want to hold a train in the upper loop.

Freight and passenger trains

It's easy to envision a Lionel N&W "J" steam locomotive blasting around this lay-out, smoke blackening the underside of the wide-span bridge at the two track diamonds, or climbing up and around the high-line reverse loop.

Scenes presented on this track plan are common throughout the Appalachian Mountains. Present-day CSX and Norfolk Southern freights will be at home on this track plan. So will their ancestors, notably the Chesapeake & Ohio, Baltimore & Ohio, Western Maryland, Virginian, Southern, and the previously mentioned N&W.

Diesel lash-ups or even twin steam locomotives pulling trains up the high-line reverse loop will be in order. Or you can run coal drags and through-service freight trains, from boxcars to modern-day double-stack shipping containers (be sure to watch your bridge clearances!).

MTH's RailKing locomotives and Lionel's LionMaster line—both slightly smaller than scale-size but with up-market detailing—will be perfect for this layout. Ditto for locomotives with speed-control electronics.

As for passenger trains, how can anyone resist the Norfolk & Western's understated Powhatan Arrow—streamliners pulled by a

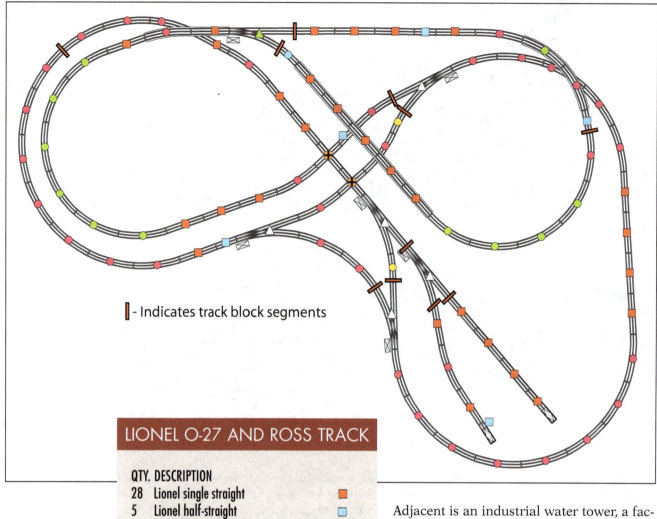

| - Indicates track block segments

LIONEL O-27 AND ROSS TRACK

QTY.	DESCRIPTION	
28	Lionel single straight	■
5	Lionel half-straight	■
13	Lionel O-42 curve (30 degrees)	●
1	Lionel O-42 half-curve (15 degrees)	●
33	Lionel O-54 curve	●
2	Lionel O-54 cut curve	●
2	Lionel 90-degree crossing	✚
2	Lionel track bumper	
1	Ross 110 O-42 right-hand turnout	▲
3	Ross 115 O-54 right-hand turnout	△
2	Ross 116 O-54 left-hand turnout	△

bullet-nosed steamer? Maybe you prefer those bright yellow C&O streamliners, led by their own sleek stainless-and-yellow streamlined steam locomotive. Both trains have been made by more than one O gauge manufacturer in scale and semi-scale sizes.

Scenery and accessories

While the suggested scenery is predominately rural, there is an industrial area on one side of the layout that includes a Lionel culvert loader and forklift platform.

Adjacent is an industrial water tower, a factory building, a new Lionel operating industrial tank, and good old Gabe the Lamplighter, a postwar American Flyer accessory now made by MTH and Lionel.

Deciduous trees and plenty of ground-level greenery dominate the interiors of the loops and cover the sloping ground that ties the upper reverse loop to the main surface of the layout.

In the loop opposite the industrial area is the suggestion of a small town, including a modern-era Lionel passenger/freight station and two MTH RailKing houses. By themselves on the far corner of the track plan are an MTH mobile home and a pair of Plasticville hobo shacks.

Finally, squeezed between track at the double diamonds is a Plasticville switch tower, selected for its compact dimensions, along with two more hobo shacks. You may be wondering: why not go for a bigger switch tower? Because there's no need to distract from that 4-foot-long series of bridges—which, as you remember, got this track plan going!

The Transcontinental

Creative toy train track plans are often a blend of ideas too numerous to count.

Rarely do just two specific sources of inspiration carry you through from

inception to completed drawings. Such is the case with this rough-and-tumble

track plan, which depicts the Central Pacific and the Union Pacific railroads

during the building of the first transcontinental railroad.

A Union Pacific train races across rugged Nevada more than a century after the first transcontinental line linked the eastern United States with the Pacific Coast.

The seeds of this winding track plan come from two specific sources: Stephen Ambrose's recent book *Nothing Like It in the World: The Men Who Built the Transcontinental Railroad*, a retelling of the story of the Central Pacific and Union Pacific railroads, and an O gauge layout story published in *Classic Toy Trains* magazine.

Portions of Ambrose's book detail the work of Chinese laborers using black powder to carve cuts and tunnels through a previously insurmountable Sierra Nevada range between Sacramento, Calif., and Reno, Nev. The task, using no mechanized machinery, was one of the greatest engineering feats of the time. Photographs taken of the work during the 1860s can only begin to describe the digging, filling, and blasting necessary to lay rails eastward through the mountains.

The December 2004 issue of *Classic Toy Trains* showcases Dick Baker's O gauge layout. His massive dog-bone-shaped layout, created by builder Clarke Dunham, features a separate mountain line that loops around itself three times.

This walkaround track plan, designed to fit within a 12- by 12-foot area, re-creates Ambrose's vision using a modified version of Baker's twisting mountain line.

Track and wiring

Starting at a tabletop level beneath a towering central bridge, a Central Pacific train on the mountain line hugs a nearly

vertical rock wall as it ascends clockwise 360 degrees not once, but twice, using O-54 and then O-42 curves. As it approaches the central bridge, it reaches the track plan's peak elevation of 12 inches.

After crossing the bridge, our train descends another 720 degrees first counterclockwise and then clockwise, looping beneath itself twice in a lazy figure-eight built from O-54 and O-42 track sections. Finally, after traveling four full rotations, our train is back at its point of origination.

The Union Pacific, which crossed the Nebraska prairie before reaching Wyoming and the Rockies, is depicted by the line of track that forms a rounded-off triangle around the base of the mountains. Connecting the two lines are four track switches and a crossing, allowing trains to move freely in and out of the mountains.

Be aware, however, that these track switches are beyond arm's reach from the front and back edges of the track plan. A wise layout builder will make the scenery inside one of the two primary loops removable to allow easy access to the track switches and the main bridge—in case Murphy's law starts to apply.

Weber Canyon, Utah, was the last geographical challenge for the Union Pacific before connecting with the tracks of the Central Pacific. Here's a spectacular shot of tunnel no. 3 on the original line. Look closely to see telegraph poles climbing the mountain above the portal. *Union Pacific Railroad*

As drawn, the track plan uses Lionel O gauge tubular track. However, you can construct the layout from other brands, although you'll need to purchase or create fitter sections near the crossing that connects the mountain line to the perimeter loop of track.

With two clearly defined lines, this track plan makes two-train operation a no-brainer. Using command control or relay-activated blocks, there's even enough room for two trains to follow one another on the outer perimeter loop, making this a three-train layout.

Conventional-control wiring is a snap. You just have to create entry and exit blocks for the mountain route, the perimeter route, the passing siding and its adjacent mainline section, and the single spur at the far end of the track plan.

Also, be sure to wire the longer of the two entry tracks into the mountains as a separate block. Then you can gracefully move one train out of the mountains and a second train into them.

A two-throttle transformer, such as a Lionel ZW, MTH Z-4000, or MRC Pure Power Dual, is ideal for this layout.

Freight and passenger trains

While 19th-century locomotives aren't typical motive power in O gauge, you do have some excellent choices. Lionel's perennial favorite General, with its quaint diamond smokestack, will look great. So will the 4-4-0 and 4-6-0 locomotives from MTH.

There are some 19th-century freight and

Who says toy train tunnels are unrealistically short? These two tunnels were photographed in 1951 on the original Central Pacific line through the Sierra Nevada Mountains. *Howard Bull*

ACCESSORIES

QTY.	DESCRIPTION
1	Atlas 6920 single truss bridge
1	Gilbert 571 truss bridge
1	Lionel 3656 operating stockyard
1	Lionel 9224 horse car platform
1	Lionel 12705 lumber shed kit
2	Lionel 12706 barrel loader building kit
2	Lionel 12711 water tower kit
1	Lionel 12734 passenger/freight station
1	Lionel 12774 lumber loader kit
2	MTH 30-90001 outhouse
1	Plasticville 1408 windmill
3	Plasticville 1627 hobo shack

passenger rolling stock on the market. MTH has produced enough different freight and passenger pieces to fill a transcontinental roster. Heck, Marx even made a diamond-stack 4-4-0 and some open platform pas-

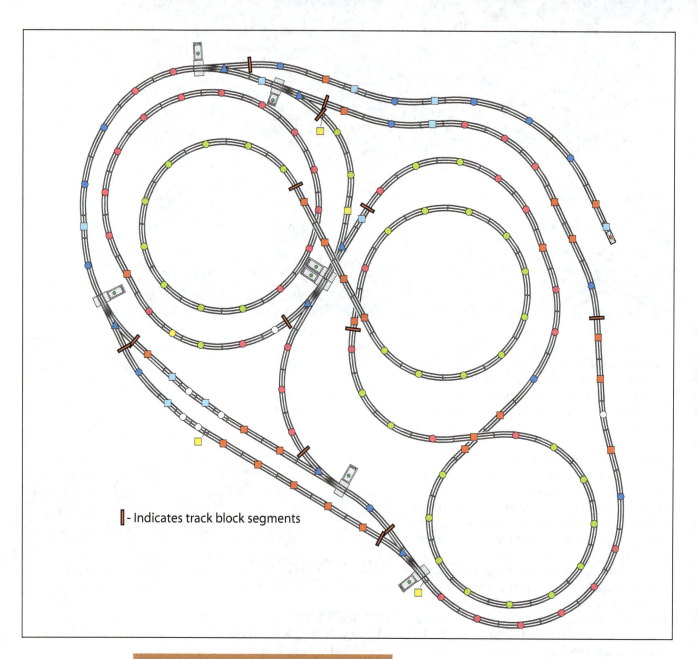

| - Indicates track block segments

LIONEL TRACK

QTY. DESCRIPTION
26   Lionel single straight   🟧
10   Lionel half-straight   🟦
4   Lionel cut straight   🟨
36   Lionel 0-42 curve (30 degrees)   🟢
1   Lionel 0-42 cut curve   🟡
37   Lionel 0-54 curve   🔴
14   Lionel 0-72 curve (22.5 degrees) 🔵
6   Lionel 0-72 cut curve   ⚪
3   Lionel 0-72 left-hand turnout   🔺
4   Lionel 0-72 right-hand turnout   🔺
1   Lionel 260 track bumper

senger coaches during the postwar years.

Compact rolling stock and short trains won't tax locomotives on the grades. The small sizes of these trains make the mountains loom that much larger.

You can also move the era you're modeling forward a few decades and run 4-6-0 Ten Wheelers and 2-8-0 Consolidations with a mix of wood-construction and metal-construction freight cars. And speaking of wood, won't Atlas O's colorful 36- and 40-foot billboard reefer cars look fantastic crawling up and down these mountains?

If teakettle locomotives aren't your cup of tea (no groans, please), this track plan lends itself to more modern trains (I can picture a pair of Geeps or smoking RS-3s double-heading at full roar up the grade),

although the contemporary Dash-9 and SD90MAC diesels may be a bit much.

Logging locomotives, such as Shays and Heislers, will be at home on the mountain as well, provided their designs can negotiate O-42 curves. K-Line released a two-truck Shay in 2004 that can negotiate curves as tight as O-31.

Scenery and accessories

Scenery should be as vertical as possible, with tall pines jutting upward from rocky outcroppings. Brush up on your rockwork, and I suggest buying one of the many "how to make scenery" books to learn rock-mold techniques. Vegetation is sparse on this track plan, so you won't be able to hide mediocre rocks with trees.

Speaking of rocks, the track plan shows boulders strewn around the perimeter of the layout. If your benchwork is sturdy, you can use appropriate natural stones to simulate 1:48 scale boulders.

There's just one tunnel on this track plan, but you can always add more. Short tunnels aren't just a toy train phenomenon—the Central Pacific really had 'em. Keeping them short adds plenty of operating interest and doesn't create problems (see the earlier Murphy's law reference).

As stated before, the 10-foot depth of the central area layout makes portions of the layout beyond arm's reach. As a result, you'll need lift-up or drop-down sections of scenery to gain access.

While scenery dominates this track plan, there is still room for accessories. In keeping with the Old West theme, a rustic modern-era Lionel freight/passenger station suffices for the small town on the front

In modern times, cargo crosses the Western mountains far more efficiently than it did in the days depicted on this track plan.
John S. Murray

edge of the layout. Along with the station are a Lionel stockyard, an MPC-era western-style water tower and plastic barrel loader, and some of the ubiquitous Plasticville hobo shacks (they seem to be at home on every track plan!).

On the backside of this track plan are another water tower and a barrel loader, an MPC-era plastic lumber loader, and a Lionel horse car platform.

The central bridge is Atlas O's 40-inch-long single truss bridge. The other bridge on the layout is an American Flyer no. 571 open truss bridge.

If either of these bridges strikes you as too modern, consider substituting hand-built wooden trestles. They're accurate alternatives and are not as tough to construct as you may think.

Do this track plan right—appropriate O gauge trains and rugged scenery—and you'll be comfortable knowing that, just as Ambrose's transcontinental book says on its cover, there's "Nothing Like It in the World."

A Day at the Races

Lots of guys just want to run trains. If the Indianapolis Motor Speedway somehow came into toy-train hands, it would be turned into the biggest three-rail oval in the world. Here's an O gauge oval-within-an-oval track plan that fulfills the need of speedway fans and then some. Two trains can run at one time, hands free. If you wire the 45-degree crossing on the outer loop for crash protection, a third train can easily be accommodated.

Bridges highlight the oval, including one of Lionel's slick new lift bridges. Off to one side of this 12- by 12-foot layout is an industrial spur highlighted by Lionel's operating culvert loader and unloader accessories.

Track and wiring

As drawn, this plan uses tubular track. Although it takes 130 pieces of track to build this layout as shown, only three pieces are cut to custom lengths—two straight sections and one O-72 curve.

The outer oval, which wraps around itself once, uses O-54 and O-72 curves. Paired O-72 track switches allow trains to enter and depart the inner oval, which uses a mix of O-42, O-54, and O-72 curves.

The inner oval is raised 4 inches above the main surface of the layout, adding interest and also justification for the two fixed bridges behind the lift bridge.

A train led by two blue Conrail diesels races across a Lionel-like girder bridge west of Toluca, Ill., in 1992. *Jeff Wilson*

The track plan you're looking at is one of the largest in this book. At its midsection, the oval is more than 8 feet across. While I've planted trees and created a river in the central areas of this plan beyond arm's reach, you'll want to construct an access hatch or two or even leave part of the center of the layout open to the floor. The raised inner oval will help mask any open floor space in the center of the layout (especially if you line the inner edge with trees and berms).

A wing extends off the oval to an indus-trial yard filled with operating accessories. The yard is 3 feet deep, so all of the accessories, even the persnickety ones, are within easy reach.

For conventional-control operation, you can make the innermost oval a separate block and block off the tracks to the left and right of the two crossovers. This arrangement will permit one train to leave the inner oval while another enters.

If you wire the 45-degree crossing on the outer oval to cut center-rail power to the

Top: A Conrail train leaves the Smith Bridge over the Hudson River near Coeymans, N.Y., in a scene similar to the bridge routes on this track plan. *Jim Conroy*

Bottom: Looking for a fast freight train for your layout? Try the Tropicana juice train, here departing the Tropicana plant in Bradenton, Fla., behind CSX power for its trip to the Northeast. *E.J. Gulash*

intersecting line when a train crosses the diamond, you'll open up the outer oval to two-train operation. Then, all you need are a handful of relay-controlled blocks to keep trains from rear-ending each other, and you're in business.

An article in the December 2004 issue of *Classic Toy Trains* magazine details two-train wiring. Another article, this time in the September 1999 issue, explains how to wire a track crossing to avoid T-bone collisions.

You'll need a two-throttled transformer for this layout. MRC's Dual Pure Power, Lionel's ZW (new or postwar) or MTH's Z-4000 will do just fine.

Freight and passenger trains

The high-speed ovals on this track plan beg for high-speed trains. Picture a Burlington Northern Santa Fe Dash-9 diesel racing across the lift bridge with a string of double-stack container cars behind (don't forget to

check the bridge height for clearance!).

Other fast freight trains can include Union Pacific refrigerator cars full of perishables heading east from the San Fernando Valley. Or how about CSX, Norfolk Southern, or Conrail boxcars carrying time-sensitive loads of auto parts to factories in the Midwest.

Even the CSX juice train—a long string of orange or white Tropicana Orange Juice reefers heading north out of Florida to New Jersey—will look right. Both Lionel and MTH have offered several different O gauge Tropicana reefers.

Passenger service is likewise fast. Pick your most colorful favorite—a Milwaukee Road Hiawatha, a Southern Pacific Daylight, or maybe even a racy purple streamliner of the Atlantic Coast Line. The choice is yours. Williams sells its O gauge locomotives and streamlined aluminum coaches in more than a dozen brilliant railroad colors.

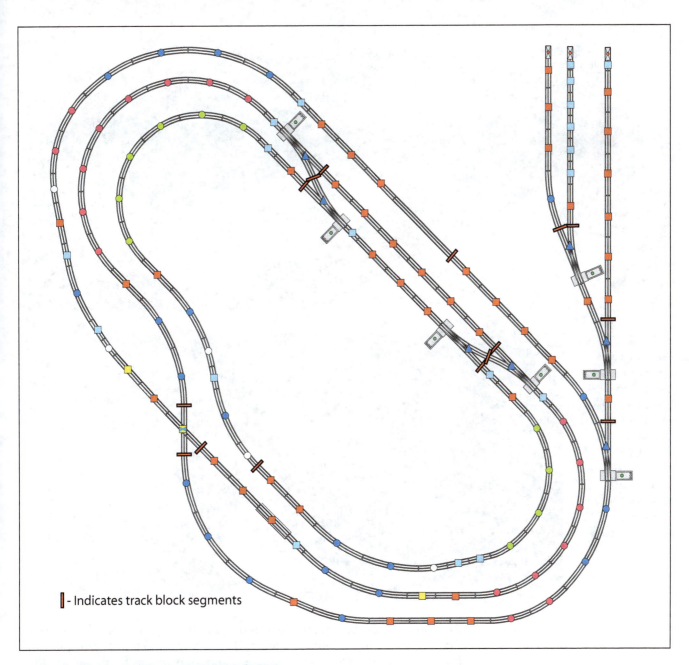

▌ - Indicates track block segments

LIONEL TRACK

QTY.	DESCRIPTION	
44	Lionel single straight	🟧
19	Lionel half-straight	🟦
2	Lionel cut straight	🟨
10	Lionel 0-42 curve (30 degrees)	🟢
18	Lionel 0-54 curve	🟠
21	Lionel 0-72 curve (22.5 degrees)	🔵
5	Lionel 0-72 cut curve	⚪
4	Lionel 0-72 left-hand turnout	🔺
3	Lionel 0-72 right-hand turnout	🔺
1	Lionel 45-degree crossing	🔲
3	Lionel 260 track bumper	

Scenery and accessories

Tracks cross the river in six different places. That necessitates three big bridges on one side of the layout and three smaller bridges on the other side.

The most appealing of these bridges is Lionel's new no. 213 lift bridge. Its cables and exposed lift mechanism—brass-colored against black girders—make this bridge a show-stopper. Behind the lift bridge is a pair of crimson MTH 30-inch truss bridges.

The river narrows as it crosses the center of the track plan. Before reaching the far side it passes below two Lionel girder bridges and an American Flyer no. 571 truss bridge.

It looks like a toy train race, but actually the Jersey Central Camelback locomotive on the left, operating at a lower speed, is simply being overtaken by the Jersey Central passenger train on the right. *Herbert Weisberger*

Elsewhere on the main section of the plan are an MTH RailKing Myerstown passenger station, an MTH RailKing switch tower, and numerous Lionel and Plasticville hobo shacks and sheds. There is a pair of Plasticville switch towers protecting the crossover tracks near the lift bridge as well as two MTH cantilever signal bridges.

Don't fret that the only operating accessory I've mentioned so far is a lift bridge. In the industrial yard are a Lionel culvert loader and unloader, a Lionel barrel loader, a modern-era Lionel operating smokestack (next to an MTH RailKing brewery), and an MTH reproduction of American Flyer's Gabe the Lamplighter tower. There's also enough room to add one or two more operating accessories.

The river is the scenic highlight of this track plan. I've also shown a pond and small tributary in the center of the layout. Depending on how you handle access hatches, you may want to duplicate this water feature.

Elsewhere in the center of the layout are a few rolling hills, lots of trees, three MTH RailKing workhouses, and a blinking water tower. Neither the trees nor the structures will need any attention once they're in place.

Because this inner part of the layout is somewhat out of reach, I've taken a conservative approach to filling it. However, you can do away with the trees and build a town if you like.

And, while you're at it, be sure to add a golf course. Huh? A golf course on your O gauge layout?

Well, if you've ever visited the Indianapolis Motor Speedway, you'll recall there's an actual golf course inside the brickyard oval!

FIFTEEN

A Twist and a Flame

In the lexicon of liquor, "with a twist" means a slice of lime or lemon on the rim of a bar glass. A slight "twist" breaks the skin of the citrus, allowing a few drops of fruit juice to cut the sometimes harsh taste of gin, vodka, or your poison of choice. In the lexicon of layouts, a "twist" is a loop of track spun around on itself, as if you picked up the end of the loop like a piece of rope and flipped it over once before setting it back down.

This L-shaped track plan, designed for two-train, hands-free operation, features an outer loop with a twist. In addition, the plan offers three spurs (one long enough to hold an entire train) and a generous passing siding, and it all fits easily within a 12- by 12-foot space.

Track and wiring

As drawn, this plan uses Ross Custom Switches track and switches. Ross rails are identical to GarGraves track rails, so the two brands, along with Curtis Hi-Rail track sections, can be freely interchanged.

Ross offers an amazing variety of track components—from slip switches to wyes to curves more than 100 inches in diameter. It's hard to imagine any layout that can't be built with Ross products.

The inner oval is more than 9 feet long and uses O-42 curves. The outer oval uses O-54 curves and includes a healthy 12-foot-long passing siding. The "twist" is a 60-degree crossing section that leads to an O-54 loop.

Follow your finger around this track plan, and you'll note that the "twist" gives a train on the outer loop a significantly

A track crossing, like this Chesapeake & Ohio diamond near Toledo, Ohio, gives this O gauge track plan its twist. *George Kleiber*

longer run than a train on the inner loop. There's no sense of "two trains chasing each other" on this track plan.

The loops are connected by three O-72 track switches and one O-42 switch. Note that there are several fitter sections of track—straight, O-42, O-54, and O-72 sections—surrounding the 60-degree crossing and the adjacent crossover switches connecting the inner and outer loops.

While this plan calls for 19 sections of

track cut to non-standard lengths, the rails of Ross/GarGraves/Curtis track are the easiest to cut of all O gauge track. Some of the modifications are as simple as cutting an O-72 curve section into two equal pieces.

Beyond the loops and the twist are a double-headed industrial spur and a 7-foot-long spur that's the perfect spot to store resting trains.

Wiring this track plan for conventional-control operation involves breaking the inner and outer loops into blocks to facilitate the movement of trains in and out of the loops. The passing siding, the adjacent section of main line, and the three spurs are blocked. Thanks to the train-storage spur, this plan can easily handle three complete trains on the layout at all times.

Freight and passenger trains

While the scenery and accessories dictate the types of trains for many of the track plans in this book, this layout's motto ought to be "any place, any time." Trains can be primarily freight or primarily passenger, and they can be pulled by modern diesels or old-time steamers.

If passenger trains are your speed, this track plan offers dual service, thanks to a passenger station with platforms on both the inner and outer loops. The outer line can be devoted to intercity streamliners, such as the aluminum O gauge beauties offered in dozens of road names by K-Line and Williams.

ACCESSORIES

QTY.	DESCRIPTION
1	Lionel 138 water tower
1	Lionel 362 barrel loader
1	Lionel 12768 burning switch tower
4	Lionel 12943 illuminated station platform
1	Lionel 14107 "497" coaling station
1	Lionel 14152 "133" passenger station
1	Lionel 22933 section gang house
1	MTH 10-1160 Lionel 165 magnetic crane
1	MTH 30-9079 townhouse 2
1	MTH 30-9080 townhouse 3
2	MTH 30-9096 CJ's Textiles factory
1	MTH 30-9112 operating firehouse
1	MTH 30-9139 house on fire
4	MTH 30-11012 operating crossing gate signal set
2	MTH 30-11014 operating crossing flasher set
1	MTH 30-90023 soda fountain corner building
1	MTH 30-90026 granary

Or how about the "American Flyer" style of New Haven Railroad coaches recently produced by Weaver? It's easy to imagine them being pulled by black-and-orange New Haven diesels or even the EP-5 electric locomotives, nicknamed "jets" for their whooshing sound. Lionel, MTH, K-Line, and Williams offer EP-5 jets.

Local or commuter service can be handled on the inner loop with heavyweight-style passenger cars, again offered by several of the major manufacturers.

On the freight front, this track plan offers three operating accessories on the

industrial spur—modern Lionel reproductions of its postwar nos. 497 coaling station and 362 barrel loader and an MTH reproduction of a Lionel no. 165 magnetic crane. Dump cars and gondolas are the freight cars of choice here.

Scenery and accessories

In addition to the barrel and coal loaders and the magnetic crane, this track plan calls for several other modern reproductions of old favorites: a Lionel no. 138 operating water tower and a no. 133 passenger station, surrounded by four illuminated station platforms.

Elsewhere on the track plan you will find MTH operating crossing signals and gates. The small downtown area comes to life with five recent-production MTH Rail-King buildings.

Scenery, too, fits the "any place, any time" design of this plan. While I've kept things simple—gravel areas around buildings and the operating accessories, trees

A New York Central Mercury streamliner kicks up dust as it rockets (at toy train-like speeds!) past the station in Michigan City, Ind., in 1947. The station resembles the one depicted on this track plan. Also note the lampposts, which are similar in style to those that are available in O gauge. *P. F. Johnson*

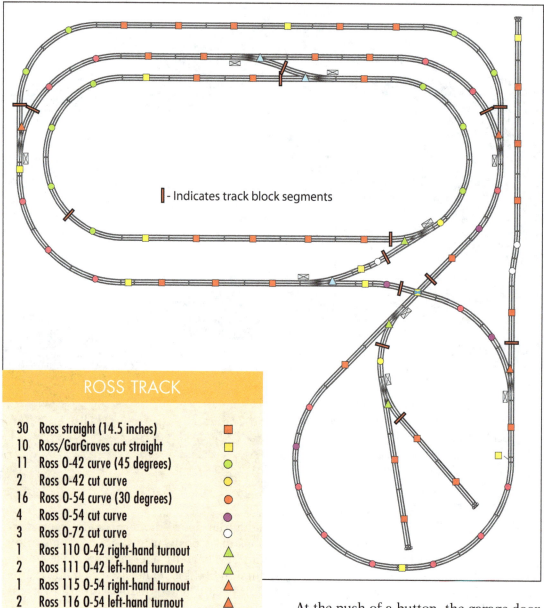

I - Indicates track block segments

ROSS TRACK

30	Ross straight (14.5 inches)	🟧
10	Ross/GarGraves cut straight	🟨
11	Ross O-42 curve (45 degrees)	🟢
2	Ross O-42 cut curve	🟡
16	Ross O-54 curve (30 degrees)	🟠
4	Ross O-54 cut curve	🟣
3	Ross O-72 cut curve	⚪
1	Ross 110 O-42 right-hand turnout	🔺
2	Ross 111 O-42 left-hand turnout	🔺
1	Ross 115 O-54 right-hand turnout	🔺
2	Ross 116 O-54 left-hand turnout	🔺
2	Ross 125 O-72 right-hand turnout	🔺
1	Ross 126 O-72 left-hand turnout	🔺
1	Ross 320 60-degree crossing	⬛
3	track bumper	

and shrubs in the flatlands, and a few roads through town—you can get as fired up as you like when building scenery. For example, I haven't specified any water features, but there are plenty of straight sections to accommodate an assortment of O gauge bridges.

Speaking of getting fired up, I've saved what may be the best part of this twist layout for last. Smack dab in the middle of the track plan is an MTH operating firehouse. This accessory rivals the MTH gasoline stations for excitement.

At the push of a button, the garage door to the firehouse raises and a die-cast metal fire truck races out of the garage bay, following a track cut into the base of the accessory. At the same time, an audio loop of firehouse sounds begins to play.

There's just one problem. While the firefighters scramble to get out of the station, there's a dilemma concerning where to go. Across the street from the firehouse is MTH's burning house accessory, complete with smoke and lighting that simulates fire. But on the other side of the firehouse is Lionel's burning switch tower, which also simulates a structure ablaze.

Whichever way the MTH fire crew turns, something is going to be left burning. Ain't that a twist—and it doesn't even involve O gauge track!

Tale of Two Cities

Here's a toy train track plan that actually goes somewhere. Yes, there are two loops for continuous running, which, when knocked down to least common denominators, are circles. But there are two parallel industrial spurs that bridge the layout and then diverge to the left and to the right. You can literally ship freight from one of the two "cities" to the other.

This shot of the Boston & Maine in Lawrence, Mass., in 1947 could have been taken on this track plan. Note the positions of the bridge, embankment, and switch tower. *Robert E. Chaffin*

Between the "cities" are two loops connected by an Atlas O double slip switch—more about this slip switch in a minute. Within the inner loop is a small town and a passenger station. The two "city" spurs began with a clever bit of trackwork called a "scissors wye." More on that in a moment, too.

Note that the industrial side of the layout has a 6-inch taller base height than the main body of the layout. While you can bring the spurs down to the layout's base level after they bridge the main line, keeping them at a uniform height prevents freight cars from rolling away in the industrial areas.

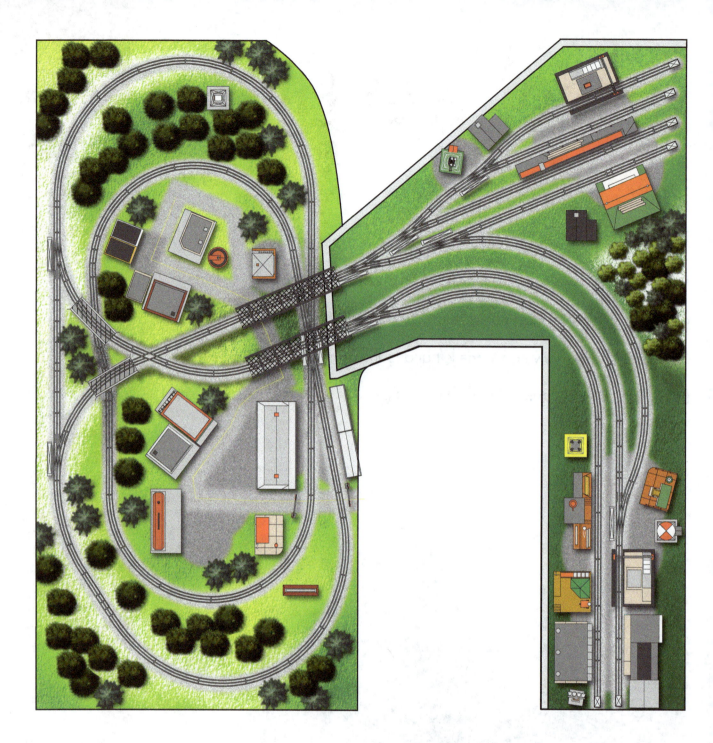

Track and wiring

As shown, this track plan uses Atlas O nickel-silver track. Atlas O offers a wide variety of specialty track sections—in this instance O-72 wye switches and a no. 5 double slip switch—that makes this track plan possible. The slip switch and the scissors wye are intriguing track features. While they set this track plan apart from others in this book, they also throw a few wrenches into what initially appears to be a routine design.

The main body of the track plan is comprised of two loops. The outer, asymmetrical loop uses curve easements—O-63 to O-54—on three of its "corners." The fourth corner is flattened a bit and comprised of three Atlas O 1.75-inch straights, an O-72 half-curve, and a full-length O-72 curve.

These fancy easements are easier on trains. They also look better and align the overall plan to accommodate the entry and departure angles of the double slip switch.

The inner loop uses Atlas O's O-45 curves. However, there's a "jog" in the back-

ACCESSORIES

QTY.	DESCRIPTION
1	Lionel 394 rotary beacon
1	Lionel 2300 oil drum loader
1	Lionel 12722 roadside diner
1	Lionel 12759 "195" floodlight tower
1	Lionel 12761 animated billboard
2	Lionel 12772 extension truss bridge
1	Lionel 12818 animated freight station
1	Lionel 12873 "464" operating sawmill
1	Lionel 12915 "164" log loader
1	Lionel 12917 "445" automatic switch tower
1	Lionel 12982 culvert loader
1	Lionel 12983 culvert unloader
1	Lionel 14001 "364" conveyor lumber loader
1	Lionel 14085 "128" animated newsstand
1	Lionel 14155 "395" floodlight tower
2	MTH 30-9006 passenger station platform

stretch (beneath the two girder bridges at the wye) to keep things aligned once the inner loop doubles back to the slip switch.

The Atlas O no. 5 double slip switch connecting the two loops is a combination of a crossing and a pair of switches. A train entering the switch from the outer loop in either direction can remain on the outer loop or, by the switch lever being thrown, enter the inner loop.

A train on the inner loop coming from either direction likewise can stay on the inner loop or cross over to the outer loop. But unlike typical crossover tracks (back-to-back track switches), the slip switch is still a crossing—two trains cannot pass through it at the same time without colliding.

Not all O gauge track manufacturers offer slip switches, so check first before deciding to build this plan with another brand of track.

The design of the scissors wye is taken straight from an Atlas O track catalog (thanks, guys!). Two O-54 track switches form the base of the wye, and two O-54 curves follow the diverging routes of the switches. On my plan, there are two half-curves on one branch to accommodate the girder bridge crossing over the jog of the inner loop.

The curves meet in a 45-degree crossing. The two spurs then continue across parallel bridges to their respective industrial areas, using four Atlas O O-72 wye track switches to create industrial yards.

The straight routes of the two O-54 switches are connected by five 4.5-inch straights and one 1.75-inch section, as specified in the Atlas O catalog. If you don't want to mess with all of that, cut Atlas O track to a total length of 24.25 inches.

On the far side of the bridges, note that the longest of the three arcing curves actually connects one "city" to the other. This gives you a few more options when switching industries. It also allows you to keep a train running on the main portion of the layout and still move cars between the freight-handling accessories.

Even if this plan isn't your cup of tea, you'll find that wye switches and slip switches can push your O gauge track plan well beyond the envelope of O-31 curves and 90-degree crossings.

Wiring for conventional-control operation calls for in-and-out blocks on the two loops to allow a train on the inner loop to swap places with one on the outer loop.

For two-train operation amid the two loops, you'll need to keep a constant eye on traffic through the slip switch and wire it using a double-throw toggle switch to allow the inner-loop throttle or the outer-loop throttle to power the rails of the switch as needed.

The two spurs are broken down into their

Top: Here's a shot of parallel truss bridges, like those depicted on this track plan. This scene didn't last long, however. The bridge on the left was being built as a replacement for the one on the right. *Bremer & Judge Associates*

Bottom: Lionel's no. 195 floodlight towers never looked better than in the Pennsylvania Railroad's Enola Yard. There are plenty of places for such lights on this layout. The loudspeaker on the roof of the yard office is used to instruct hump-track workers. *Philip R. Hastings*

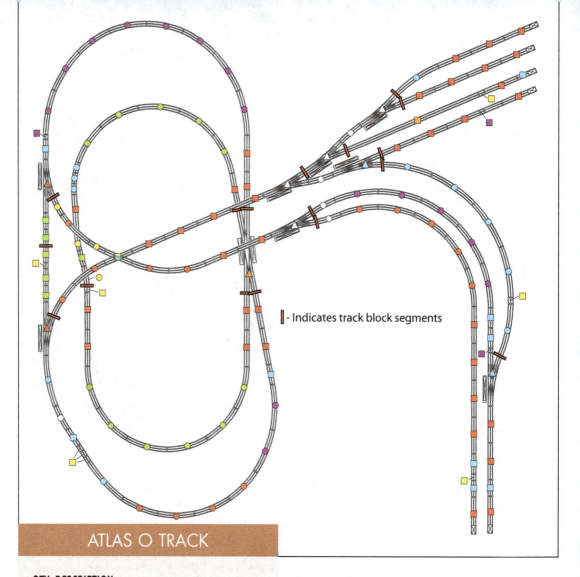

- Indicates track block segments

ATLAS O TRACK

QTY.	DESCRIPTION	
3	Atlas 6015 1.25-inch straight	■
1	Atlas 6027 no. 5 double slip switch	▲
6	Atlas 6040 snap-on bumper	
11	Atlas 6045 O-45 curve (30 degrees)	●
3	Atlas 6046 O-45 curve (7.5 degrees)	●
36	Atlas 6050 10-inch straight	■
8	Atlas 6051 4.5-inch straight	■
8	Atlas 6052 1.75-inch straight	■
8	Atlas 6053 5.5-inch straight	■
1	Atlas 6058 40-inch straight	■
18	Atlas 6060 O-54 curve (22.5 degrees)	●
4	Atlas 6061 O-54 curve (11.25 degrees)	●
8	Atlas 6062 O-72 curve (22.5 degrees)	●
5	Atlas 6063 O-72 curve (11.25 degrees)	○
11	Atlas 6064 O-63 curve (22.5 degrees)	●
1	Atlas 6070 O-54 left-hand turnout	▲
1	Atlas 6071 O-54 right-hand turnout	▲
2	Atlas 6073 O-72 right-hand turnout	▲
4	Atlas 6074 O-72 wye	△
1	Atlas 6082 45-degree crossing	▣

basic components to facilitate train movement on one line without power on the adjacent lines.

Freight and passenger trains

This track plan is all about freight cars moving goods back and forth between "city" spurs.

There are two log-loading accessories along one city spur and a Lionel operating sawmill at the other. Building this track plan as depicted calls for several log-hauling flatcars. Ditto for gondolas to bring Lionel culverts from the loader accessory to the unloader. Add more gondolas for the oil-drum loader and assorted boxcars to service the MTH RailKing warehouse and operating transfer dock.

You can haul these cars with diesel or steam locomotives. The look of this track plan says transition-era, when the last steam locomotives and the first generation of diesels shared the rails. Don't forget to

keep a few switching locomotives on your roster—those operating front couplers will be handy on a layout built from this plan.

While the focus is on the movement of freight, there's a healthy-sized passenger station servicing both the main and inner loops. Train service is your choice: speedy streamliners or local service with heavyweight coaches or Rail Diesel Cars.

Scenery and accessories

As mentioned, the two city spurs are full of postwar Lionel accessories and their modern-era counterparts. There's also plenty of space for an MTH RailKing operating transfer dock, a modern-era Lionel animated freight station, and reproductions of American Flyer's Gabe the Lamplighter and water tower.

Within the main loops are a town using MTH RailKing buildings and a two-story passenger station with platforms. Nearby are Lionel's smoking roadside diner, a modern-era reproduction of Lionel's animated switch tower, and an animated billboard. The two big truss bridges spanning the slip switch are from Atlas O. Capping the accessories is an MTH American flag.

Scenery consists of gentle hills blending the outer loop with the inner loop as it climbs and then descends at the scissors wye. Both spurs and the 45-degree crossing form an elevated line that cuts across the center of the two loops, necessitating about 10 linear feet of retaining wall. Check the Scenic Express catalog or website (www.scenicexpress.com) for plenty of retaining-wall choices.

Beyond that, trees, other greenery, and a road through town with a pair of operating crossing gates complete the layout. If your two city spurs are against a wall of your train room, separate urban backdrops will add a final exclamation mark to this "tale of two cities" track plan.

Conclusion

If you've arrived at this page in *Creative Toy Train Track Plans*, you've not only reached the end of the book, but you very well may be at the starting point of your next toy train layout.

Use one of these 16 track plans as presented to build your dream layout; merge the features of one plan with those of another; change the scenery or the dimensions of the layout; or start with a perfectly clean sheet of paper and the inspiration you gathered from reading this book. It's all up to you.

While track styles, scenery techniques, and the very trains we operate on our layouts will continue to grow and evolve, a successful track plan will stand the test of time and lead to years of enjoyment.

Good luck on your next layout, and should you build your empire using one of these plans, by all means let the staff at *Classic Toy Trains* magazine know. Now put this book down and get started!

LIST OF MANUFACTURERS

Atlas O LLC
378 Florence Ave.
Hillside, NJ 07205
www.atlasO.com
908-687-9590

Curtis Hi-Rail Products
P.O. Box 385
North Stonington, CT 06359
curtishirail.ce.net (no www)
800-277-7245

GarGraves Trackage Corp.
8967 Ridge Rd.
North Rose, NY 14516
www.gargraves.com
315-483-6577

K-Line Electric Trains
P.O. Box 2831
Chapel Hill, NC 27515
www.k-linetrains.com
919-929-8420

Lionel LLC
50625 Richard W. Blvd.
Chesterfield, MI 48051
www.lionel.com
586-949-4100

MTH Electric Trains
7020 Columbia Gateway Dr.
Columbia, MD 21046
www.mth-railking.com
410-381-2580

R & S Enterprises
RR-Track software
P.O. Box 643
Dept. RRT
Jonestown, PA 17038
www.rrtrack.com
717-865-3444

Ross Custom Switches
45 Church St.
Norwich, CT 06360
www.rossswitches.com
800-331-1395

Williams Electric Trains
8835-F Columbia 100 Pkwy.
Columbia, MD 21045
www.williamstrains.com
410-997-7766

CPSIA information can be obtained
at www.ICGtesting.com
Printed in the USA
LVHW010335191120
672010LV00010B/567

9 780897 785303